STUDENT MATHEMATICAL LIBRARY
Volume 19

Computable Functions

A. Shen
N. K. Vereshchagin

Translated by
V. N. Dubrovskii

AMERICAN MATHEMATICAL SOCIETY

Н. К. Верещагин, А. Шень

ВЫЧИСЛИМЫЕ ФУНКЦИИ

МЦНМО, 1999

Translated from the Russian by V. N. Dubrovskii

2000 *Mathematics Subject Classification.* Primary 03-01, 03Dxx.

Library of Congress Cataloging-in-Publication Data

Vereshchagin, Nikolai Konstantinovich, 1958–
   Computable functions / A. Shen, N.K. Vereshchagin ; translated by V.N. Dubrovskii.
      p. cm. — (Student mathematical library, ISSN 1520-9121 ; v. 19)
   Authors' names on t.p. of translation reversed from original.
   Includes bibliographical references and index.
   ISBN 0-8218-2732-4 (alk. paper)
   1. Computable functions.   I. Shen, A. (Alexander), 1958–   II. Title.   III. Series.

QA9.59.V47   2003
511.3—dc21
                                                    2002038567

Contents

Preface

This book is based on lectures for undergraduate students given by the authors at the Moscow State University Mathematics Department. (Another book by the authors, "Basic Set Theory", was published as volume 17 in this book series.)

The theory of computable functions appeared in the 1930s, when no computers (in the modern sense of the word) existed. The first computers were designed in the 1940s, and one of their designers was the English mathematician Alan Turing, one of the creators of the theory of computable functions. In 1936 he described the abstract machines later called *Turing machines*; supposedly, the idea of a program stored in a computer's memory was inspired by the proof of the existence of a universal Turing machine.

For this reason alone, the basic notions of computability theory (the general theory of algorithms) deserve the attention of mathematicians and programmers. But this theory also has wider cultural significance. In 1944 one of its founders, the American mathematician Emil Post, wrote that in the history of discrete mathematics, the formulation of the notion of computability can play a role second in importance only to the formulation of the notion of the natural number.

Perhaps, nowadays, Post's statement seems to be an exaggeration: in the last decades it became clear that, philosophically, the difference between feasible and unfeasible problems is no less essential than the difference between decidable and undecidable problems, and computational complexity theory took up one of the central positions in logic, mathematics, and computer science.

Computational complexity is beyond the scope of this book. Our goal is restricted: we attempt to select central notions and facts from the general theory of algorithms and present them clearly, trying not to shadow simple general ideas by technical details. We do not assume any special preliminary knowledge, although we assume a certain level of mathematical culture (and do not explain, say, what a function or a real number is).

We hope that the reader will enjoy this first acquaintance with the theory of algorithms. To learn more about this theory (which is the central part of mathematical logic), the reader is referred to the books listed in the Bibliography.

The authors thank their teacher Vladimir Andreevich Uspensky, whose lectures, texts, and opinions, perhaps, had an even greater effect on them (and on the contents of this book) than they are aware of.

We thank the staff and all the students of the Division of Mathematical Logic and the Theory of Algorithms (Moscow State University, Mathematics Department), as well as all participants of our lectures and seminars, and readers of the draft versions of this book.

Finally, we are grateful to the American Mathematical Society and Sergei Gelfand, who organized the English translation of this book, and to Vladimir Dubrovsky, who translated it with great care.

The authors will be grateful for any reports on errors and typos (e-mail: `ver@mccme.ru`, `shen@mccme.ru`; address: Moscow Center of Continuous Mathematical Education, Bolshoi Vlasyevskii per. 11, Moscow, Russia 121002).

N. K. Vereshchagin, A. Shen

Chapter 1

Computable Functions, Decidable and Enumerable Sets

1. Computable functions

A function f with natural arguments and values is called *computable* if there exists an algorithm that computes f, that is, an algorithm A such that

- if $f(n)$ is defined for a certain natural n, then the algorithm A halts on the input n and prints $f(n)$;

- if $f(n)$ is undefined, then the algorithm A does not halt on the input n.

A few remarks concerning this definition.

1. The notion of computability is defined for *partial* functions (whose domain is a subset of natural numbers). For instance, the *empty* (*nowhere defined*) function (that is, the function with empty domain) is computable: consider an algorithm A that never terminates no matter what the input is.

2. We could have modified the definition by saying, "if $f(n)$ is undefined, then either the algorithm A does not halt or it halts but does not print anything". In fact, this would not really change anything (instead of halting without any output, the algorithm can go into an infinite loop).

3. To avoid misunderstanding, we should explain that by *natural numbers* we mean nonnegative integers (rather than only positive as is often the case). It is clear that our definition of computable functions does not depend on the specific representation of natural numbers (an algorithm that deals, say, with binary representation can be easily modified to accept and produce decimal representations, etc.). Instead of natural numbers, algorithms can have as their input and output binary strings ("words" in the alphabet $\{0, 1\}$), pairs of natural numbers, finite sequences of strings, and, in general, any "constructive objects". Therefore, we can similarly define, say, the notion of a computable function of two natural variables whose values are rational numbers.

Note that the case of functions with real arguments and values is more difficult. The computability of a function $f: \mathbb{R} \to \mathbb{R}$ needs a special definition; moreover, this definition can be given in several ways. We are not going to discuss the computability of such functions. Let us only notice that, for instance, the sine function is computable (under a reasonable definition of computability), whereas the $\text{sign}(x)$ function, equal to -1, 0, and 1 for $x < 0$, $x = 0$, and $x > 0$, respectively, is not.

As well, a special definition is required for the computability of functions whose arguments are infinite sequences of zeros and ones, etc.

4. A few decades ago the notion of algorithm had to be carefully explained. Nowadays (due to "computer literacy") nobody will read such explanations anyway. You can think about algorithms as programs in your favorite programming language. You should only assume that the memory is unlimited and the integers in use (array indices, pointers, etc.) are unbounded. It is clear that the notion of computability does not depend on the language: to port, say, a C++ program to C could be a tedious task, but is always possible theoretically.

However, you should be careful not to mistake for an algorithm something which is not an algorithm. Here is an example of a false reasoning.

Let us "prove" that any computable function f with natural arguments and values can be extended to a *total* (i.e., defined on the entire set $\mathbb{N}$ of natural numbers) computable function $g\colon \mathbb{N} \to \mathbb{N}$. Indeed, if f is computed by an algorithm A, then the following algorithm B computes a total function g extending f: "if A halts on n, then B yields the same result as A; if A does not halt on n, then B returns 0". (What is wrong with this argument?)

2. Decidable sets

A set X of natural numbers is called *decidable* if there exists an algorithm that determines whether an arbitrarily given natural number n belongs to the set X. Such an algorithm must terminate for any n and give one of the two answers "yes" or "no" (or 1/0, TRUE/FALSE, etc.)

In other words, X is decidable if its *characteristic function*

$$\chi(n) = (\textbf{if } n \in X \textbf{ then } 1 \textbf{ else } 0 \textbf{ fi})$$

is computable.

Obviously, the intersection, union, and set difference of decidable sets are decidable. Any finite set is decidable.

The decidability of sets of pairs of natural numbers, sets of rational numbers, etc. is defined similarly.

Problem 1. Prove that the set of all rational numbers smaller than the number e (the base of the natural logarithm) is decidable.

Problem 2. Prove that a nonempty set of natural numbers is decidable if and only if it is the range of a total nondecreasing computable function with natural arguments and values.

Let us mention one subtle point: the decidability of a set can be proved nonconstructively, without any explicit description of a deciding algorithm. A traditional example is the set of all n for which the number π contains at least n nines in a row. This set is decidable, because it consists either of all natural numbers (and therefore is decidable) or of all natural numbers up to a certain one (and is decidable as any finite set). So we have proved that this set is decidable in any case. However, we have not explicitly provided an algorithm that could determine, for a given n, whether the number π contains at least n nines in a row.

Problem 3. Have we used any properties of the number π in this argument? What will change if we replace the words "at least n nines" by "exactly n nines (surrounded by non-nines)"?

Do undecidable sets exist? The answer is obviously positive, because there are countably many algorithms (and therefore, countably many decidable subsets of $\mathbb{N}$), whereas the set of all subsets of $\mathbb{N}$ is uncountable. Concrete examples will be constructed later.

3. Enumerable sets

A set of natural numbers is called *enumerable* if it can be enumerated by a certain algorithm, that is, if there exists an algorithm that prints (in an arbitrary order and with arbitrary delays) all elements of this set and only them.

Such an algorithm has no input; having printed several numbers it can abruptly sink into a lengthy computation and print the next number only after a lapse of time or never print anything at all (this means that the set is finite).

There are many other equivalent definitions of an enumerable set. Here are some of them:

(1) A set is enumerable if it is the domain of a computable function.

(2) A set is enumerable if it is the range of a computable function.

(3) A set X is enumerable if its (as they sometimes say) *semicharacteristic* function, which is, by definition, equal to 0 on elements of X and undefined outside X, is computable.

Let us prove the equivalence of all the definitions above; in the proof, (0) will stand for our initial definition.

$(0) \Rightarrow (1), (3)$ Suppose that X is enumerated by an algorithm A. Then the semicharacteristic function of the set X is computable. Indeed, it is computed by the following algorithm:

> Having received a number n at the input, execute the algorithm A step by step waiting until the number n is printed. As soon as it is, send 0 to the output and terminate.

$(1) \Rightarrow (0)$ Let X be the domain of a (computable) function f computed by a certain algorithm B. Then X can be enumerated by the following algorithm A:

> Execute algorithm B step by step in parallel on the inputs $0, 1, 2, \ldots$ gradually increasing the number of involved inputs and the number of steps performed (first, one step of B is performed on the inputs 0 and 1; then two steps are performed on each of the inputs $0, 1, 2$; then three steps are performed on each of the inputs $0, 1, 2, 3$, and so on). All the arguments on which the algorithm B terminates are printed out as soon as they are detected.

Since definition (3) obviously implies definition (1), the equivalence of these definitions to the original one has been established.

To prove that $(2) \Rightarrow (1)$, i.e., to obtain an algorithm that enumerates the range of a computable function f, we only have to modify the algorithm A described above so as to make it print the results returned by B rather than the arguments on which B terminates.

It remains to prove that $(1) \Rightarrow (2)$, i.e., to show that any enumerable set X is the range of a computable function. We already know that X is the domain of a computable function. If this function is computed by an algorithm A, then X is the range of the function b that takes the value x if A terminates on the input x and is undefined otherwise. We will write this and similar function definitions as

$$b(x) = \begin{cases} x \text{ if } A \text{ terminates on } x, \\ \text{undefined otherwise.} \end{cases}$$

The algorithm that computes this function operates exactly like A, but instead of the result produced by the algorithm A it prints out a copy of the input.

Here is one more equivalent definition of an enumerable set: a set X of natural numbers is enumerable if X either is empty or is the range of a total computable function (in other words, its elements can be arranged into a computable sequence).

Indeed, suppose that a nonempty enumerable set X is enumerated by an algorithm A. Let x_0 be an arbitrary element of X. Consider the following total function a: if A returns a number t at the nth step, then $a(n) = t$; if nothing is returned, then $a(n) = x_0$. (We assume that only one number can appear at any given step; otherwise the computation must be split into smaller steps.)

It should be mentioned that this reasoning is nonconstructive. That is, given an algorithm A, we do not necessarily know whether the set it enumerates is empty or not.

Theorem 1. *The intersection and union of enumerable sets are enumerable.*

Proof. If X and Y are enumerated by algorithms A and B, then their union is enumerated by the algorithm that performs A and B in parallel and prints everything printed by A and B. The case of intersection is a bit more difficult: the results produced by A and B must be stored and compared; common results are printed.   □

Problem 4. Prove Theorem 1 using one of the other equivalent definitions of enumerability.

As we will see, the complement of an enumerable set may not be enumerable.

Problem 5. Sometimes "nondeterministic algorithms" are considered (this oxymoron is used fairly often). Such an algorithm involves instructions like

n := arbitrary natural number

(the instruction "n := 0 or 1" will suffice, though, because any number can be composed bit by bit). A nondeterministic algorithm can follow different computation paths for the same input, depending on the "arbitrary numbers" it chooses. Prove that an enumerable set can be equivalently defined as the set of numbers that can appear at the output of a nondeterministic algorithm (with a fixed input).

Problem 6. Prove that if sets $A \subset \mathbb{N}$ and $B \subset \mathbb{N}$ are enumerable, then their Cartesian product $A \times B \subset \mathbb{N} \times \mathbb{N}$ is also enumerable.

4. Enumerable and decidable sets

Theorem 2. *Any decidable set of natural numbers is enumerable. If a set A and its complement $\mathbb{N} \setminus A$ are enumerable, then A is decidable.*

Proof. If there is an algorithm that tests whether a number belongs to a set A, then A and its complement are enumerable: for each of the numbers $0, 1, 2, \ldots$, we check if it belongs to A and print the numbers that do (or those that do not).

Conversely, assume that we have an algorithm enumerating A and another algorithm enumerating the complement of A. Then to find out whether a given number n belongs to A we run both algorithms and wait until one of them prints n (we know that exactly one of them will print n eventually). Then we check which of the algorithms has printed the number, and thus find out whether n belongs to A or not. □

This fact is called *Post's Theorem.*

It says that decidable sets are the enumerable sets with enumerable complements. On the other hand, enumerable sets can be defined in terms of decidability.

Theorem 3. *A set P of natural numbers is enumerable if and only if P is the projection of a decidable set Q of pairs of natural numbers. (By the projection of a set of pairs we mean the set of their first components: $x \in P \Leftrightarrow \exists y (\langle x, y \rangle \in Q)$.)*

Proof. The projection of any enumerable set of pairs is enumerable (we enumerate the pairs and extract their first elements), so the projection of a decidable set is all the more enumerable.

Conversely, an enumerable set P enumerated by an algorithm A is the projection of the decidable set Q consisting of all the pairs $\langle x, n \rangle$ whose first element x appears during the first n steps performed by the algorithm A. (This property of a pair $\langle x, n \rangle$ is obviously decidable.)

□

5. Enumerability and computability

We have seen that the notion of an enumerable set can be defined in terms of computable functions (for instance, as the domain of a computable function). This situation can be inverted.

Theorem 4. *A function f with natural arguments and values is computable if and only if its graph*

$$F = \{\langle x, y \rangle \mid f(x) \text{ is defined and equal to } y\}$$

is an enumerable set of pairs of natural numbers.

Proof. Suppose that f is computable. Then there exists an algorithm that enumerates its domain, that is, prints all the x at which the function f is defined. By adding the computation of the value $f(x)$ for each of these x, we get an algorithm that enumerates the set F.

Conversely, if there is an algorithm A enumerating F, then the function f is computed by the following algorithm: given a number n at the input, run the algorithm A and wait until it returns a pair with the first component n; as soon as this happens, print the second component of the pair and terminate the computation.   □

Let f be a partial function with natural arguments and values, and let A be a subset of $\mathbb{N}$. The *image* of the set A under f is defined as the set of all the numbers $f(n)$ such that $n \in A$ and $f(n)$ is defined. The *preimage* of the set A under f is defined as the set of all n such that $f(n)$ is defined and belongs to A.

Theorem 5. *The image and preimage of an enumerable set under a computable function are enumerable.*

Proof. Indeed, to obtain the preimage of an enumerable set A under a computable function f, it suffices to intersect the graph of f with the enumerable set $\mathbb{N} \times A$ and take the projection of the intersection on the first coordinate. A similar argument, with coordinates swapped, applies to the image.   □

Problem 7. Let F be an enumerable set of pairs of natural numbers. Prove that there exists a computable function f defined on the set $\pi_1(F) = \{x \mid \exists y \, \langle x, y \rangle \in F\}$, such that $\langle x, f(x) \rangle \in F$ for any $x \in \pi_1(F)$. (Sometimes this statement is called the *Uniformization Theorem*.)

Problem 8. Let X and Y be two enumerable sets with nonempty intersection. Prove that there exist disjoint enumerable sets $X' \subset X$ and $Y' \subset Y$ such that $X' \cup Y' = X \cup Y$.

Problem 9. A *Diophantine* equation is an equation of the form $P(x_1, \ldots, x_n) = 0$, where P is a polynomial with integer coefficients. Prove that the set of Diophantine equations that have integer solutions is enumerable. (This set is undecidable as was shown by Yu. V. Matiyasevich in his well-known solution of the famous "10th Hilbert Problem".)

Problem 10. Without referring to the proof of the Fermat Great Theorem, show that the set of all natural n such that the equation $x^n + y^n = z^n$ has positive integer solutions is enumerable. (As we know now, this set contains only the numbers 1 and 2.)

Problem 11. Show that any infinite enumerable set can be represented in the form $\{a(0), a(1), a(2), \ldots\}$, where a is a total computable function with pairwise distinct values. (*Hint*: delete repetitions in the course of enumeration.)

Problem 12. Show that any infinite enumerable set contains an infinite decidable subset. (*Hint*: use the previous problem and choose an increasing subsequence.)

Problem 13. Show that for any computable function f, there exists a computable function which is "pseudoinverse" to f in the following sense: the domain of g coincides with the range of f and $f(g(f(x))) = f(x)$ for all x such that $f(x)$ is defined.

Problem 14. A real number α is called *computable* if there exists a computable function a which, for any rational $\varepsilon > 0$, yields a rational approximation to α accurate to ε, i.e., $|\alpha - a(\varepsilon)| \le \varepsilon$ for any rational $\varepsilon > 0$. (Rational numbers are constructive objects, so the notion of computability does not need any special refinement.)

(a) Prove that α is computable if and only if the set $\{q \in \mathbb{Q} \mid q < \alpha\}$ is decidable.

(b) Prove that α is computable if and only if the digits of the decimal (or binary) fraction representing it form a computable sequence.

(c) Prove that the number α is computable if and only if there exists a computable sequence of rational numbers computably converging to α (computable convergence means that, for any ε, the corresponding number N in the standard ε-N definition can be found algorithmically).

(d) Show that the sum, product, difference, and ratio of computable real numbers are computable. Show that roots of a polynomial with computable coefficients are computable.

(e) What is a computable sequence of computable real numbers? A computably convergent sequence of real numbers? (Give natural definitions.) Prove that the limit of a computably convergent sequence of computable real numbers is a computable real number.

(f) A real number α is called *enumerable from below* if the set of all rational numbers smaller than α is enumerable. (Real numbers enumerable from above are defined similarly.) Prove that a number α is enumerable from below if and only if it is the limit of a computable increasing sequence of rational numbers.

(g) Prove that a real number is computable if and only if it is enumerable both from below and from above.

For further properties of computable real numbers see Problem 23.

Chapter 2

Universal Functions and Undecidability

1. Universal functions

Now we will construct a set that is enumerable but not decidable. To do this, we will use the notion of a universal function.

A function U of two natural variables[1] is said to be *universal* for the class of all computable functions of one variable if

(1) for any n, the function

$$U_n \colon x \mapsto U(n, x)$$

(the *section* of the function U for the chosen n) is computable;

(2) all computable functions of one variable occur among the sections U_n.

(Recall that neither the function U nor the one-variable functions need be total.)

A similar definition can be given for other classes of unary functions: for instance, a binary function U is called universal for the class

---

[1]Shorter terms for functions of one, two, three, and, in general, k variables are *unary*, *binary*, *ternary*, and *k-ary* functions, respectively.

of total computable unary functions if all its sections U_n are total computable unary functions and any such function appears among U_n. Obviously, a universal function exists for any countable class (and only for these classes).

The following fact plays a key role in this chapter.

Theorem 6. *There exists a binary computable function universal for the class of unary computable functions.*

Proof. Let us choose some programming language and arrange all the programs that compute unary functions into a computable sequence $p_0, p_1, \ldots$ (for instance, in ascending order of their length). Set $U(i, x)$ to be equal to the output of the ith program run on the input x. Then the function U is just the desired computable universal function. The section U_i is the computable function computed by the program p_i. In essence, the algorithm computing the function U itself is an interpreter for the programming language we use. Indeed, an interpreter for a given programming language takes two arguments—a program p in this language and an input x—and simulates the behavior of p on x. It is well known that an interpreter of some programming language can be programmed in the same language (for example, an interpreter for Pascal programs can be written in Pascal). Identifying a program with its number, we can say that a universal function is an interpreter that applies its first argument to the second one. □

Problem 15. Assume that all the sections U_n of a binary function U are computable. Does it imply that the function U is computable?

Problem 16. Give a (natural) definition of a ternary computable function universal for the class of binary computable functions and prove that such a function exists.

A similar terminology is used for sets: a set $W \subset \mathbb{N} \times \mathbb{N}$ is called *universal* for a certain class of sets of natural numbers if all the sections

$$W_n = \{ x \mid \langle n, x \rangle \in W \}$$

of the set W belong to this class and there are no other sets in it.

Theorem 7. *There exists an enumerable set of pairs of natural numbers universal for the class of all enumerable sets of natural numbers.*

Proof. Consider the domain of a universal function U. It is a universal enumerable set, because any enumerable set is the domain of some unary computable function, and, therefore, of the function U_n for some n. □

Problem 17. How can a universal set be constructed based on the fact that any enumerable set is the range of a certain function U_n?

Problem 18. Does there exist a decidable set of pairs of natural numbers that is universal for the class of all decidable sets of natural numbers?

2. The diagonal construction

In the previous section we constructed a computable universal function for the class of all computable functions of one variable. Is it possible to do the same for the class of all total computable unary functions? It turns out that this is impossible.

Theorem 8. *There is no total computable function of two variables universal for the class of all total computable functions of one variable.*

Proof. We will apply the "diagonal construction"; a similar idea is used to prove that the set of all infinite decimal fractions is uncountable (see, e.g., [**12**, Chapter 1, §6]). Let U be an arbitrary total computable function of two variables. Consider the diagonal function $u(n) = U(n, n)$. Obviously, on the argument n, the function u coincides with the function U_n (i.e., $u(n) = U_n(n)$), and $d(n) = u(n)+1$ differs from $U_n(n)$. Hence the total computable function d differs from all the sections U_n, and so the function U is not universal. □

Why does this argument not work for the class of all computable functions (including partial functions)? The point is that in this case the value $d(n) = U(n, n)+1$ is not necessarily distinct from the value $U_n(n) = U(n, n)$ since both values can be undefined.

However, a part of our argument remains valid.

Theorem 9. *There exists a computable function d such that no computable function f (with natural arguments and values) can differ from it everywhere: for any computable unary function f there exists a number n such that $f(n) = d(n)$ (this equation implies that either both values $f(n)$ and $d(n)$ are undefined or they both are defined and equal).*

Proof. The desired function d is the diagonal function, i.e., $d(n) = U(n,n)$ (where U is a binary computable function universal for the class of unary computable functions). Any computable function f coincides with U_n for a certain n; hence $f(n) = U_n(n) = U(n,n) = d(n)$. $\qquad\square$

Theorem 10. *There exists a computable function that has no total computable extension.*

Proof. The desired function can be defined by the formula $d'(n) = d(n) + 1$, where d is the function from the previous theorem. Indeed, any of its total extensions $\overline{d'}$ differs from d everywhere (if $d(n)$ is defined, then $d'(n) = d(n) + 1$ is defined and $\overline{d'}(n) = d'(n) \neq d(n)$; if $d(n)$ is undefined, then $\overline{d'}(n) \neq d(n)$, since $\overline{d'}$ is a total function), and, therefore, $\overline{d'}$ is not computable. $\qquad\square$

Problem 19. Prove that the function d (Theorem 9) itself has no total computable extension either.

3. Enumerable undecidable set

Now we can fulfill the promise we gave at the beginning of this chapter.

Theorem 11. *There exists an enumerable undecidable set of natural numbers. (Restatement: there exists an enumerable set with nonenumerable complement.)*

Proof. Consider a computable function f with natural arguments and values that has no total computable extension. Its domain F is the desired set. Indeed, F is enumerable (by one of the definitions of enumerability). If F were decidable, then the function

$$g(x) = \begin{cases} f(x) & \text{if } x \in F, \\ 0 & \text{if } x \notin F \end{cases}$$

would be a total computable extension of the function f (to compute $g(x)$, we check if x belongs to F (we can do this since F is decidable), and if $x \in F$, we compute $f(x)$). □

It is instructive to trace the construction back and look closer at the set F that eventually turned out to be enumerable and undecidable. In our construction F is the set of all n such that $U(n, n)$ is defined. Recalling the construction of the universal function U, we see that F is the set of all n such that the nth program halts on n. One can say that the problem "find out whether a given program terminates when applied to its own number" is algorithmically unsolvable.

Therefore, a more general problem, "for a given algorithm A and input x, find out whether A halts on x", is undecidable. This is the famous "Halting Problem".

In other terms, the domain of the function U is also an enumerable undecidable set of pairs. (An algorithm that tells whether a program terminates when applied to any input can also tell whether a program terminates when applied to its own number.)

Problem 20. Let U be any enumerable set of pairs of natural numbers that is universal for the class of all enumerable sets of natural numbers. Prove that its "diagonal section" $K = \{x \mid \langle x, x \rangle \in U\}$ is an enumerable undecidable set.

Problem 21. Let a set S of natural numbers be decidable. We factor all the numbers from S into primes and form the set D of all the prime numbers that occur in these factorizations. Is the set D always decidable?

Problem 22. Let a set $U \subset \mathbb{N} \times \mathbb{N}$ be decidable. Is it true that the set of "lower points" of the set U, that is, the set

$$V = \{\langle x, y \rangle \mid (\langle x, y \rangle \in U) \text{ and } (\langle x, z \rangle \notin U \text{ for all } z < y)\},$$

is decidable? Is it true that V is enumerable if U is enumerable?

Problem 23. Show that there exists a real number α that is enumerable from below, but not computable (see the definitions on pp. 9 and 10). (*Hint.* Consider the sum of the series $\sum 2^{-k}$ over all the k from an enumerable set P. This sum is always enumerable from below, but it is computable only if P is decidable.)

We return to computable real numbers in Problems 31 and 60.

4. Enumerable inseparable sets

A minor modification of our reasoning enables us to improve the result obtained above.

Theorem 12. *There exists a computable function that takes only the values 0 and 1 and has no total computable extension.*

Proof. Instead of the function $d'(x) = d(x) + 1$ we can consider the function

$$d''(x) = \begin{cases} 1 \text{ if } d(x) = 0, \\ 0 \text{ if } d(x) > 0 \end{cases}$$

(this equation implicitly assumes that $d''(x)$ is undefined if $d(x)$ is undefined). Then any total extension of the function d'' differs from d everywhere, as before, and so is not computable. □

This result can be translated into the language of enumerable sets. We say that a set C *separates* two disjoint sets X and Y if C contains one of them and has no common points with the other.

Theorem 13. *There exist two disjoint enumerable sets X and Y that cannot be separated by any decidable set.*

Proof. Indeed, let d be a computable function that takes only the values 0 and 1 and has no total computable extension. Set $X = \{x \mid d(x) = 1\}$ and $Y = \{x \mid d(x) = 0\}$. It is easy to see that the sets X and Y are enumerable. Suppose that they can be separated by a decidable set C; we may assume that C contains X and is disjoint with Y (otherwise, consider the complement of C). Then the characteristic function of the set C (equal to 1 on C and to 0 outside C) extends d. □

Notice that Theorem 13 implies the existence of an enumerable undecidable set (if two sets cannot be separated by a decidable set, then none of them is decidable).

Problem 24. How can the enumerable inseparable sets constructed above be described in terms of the universal function $U(n, x)$?

Problem 25. Show that there exist countably many disjoint enumerable sets such that any two of them are inseparable (cannot be separated by a decidable set).

5. Simple sets: The Post construction

There are other constructions of enumerable undecidable sets. One of them (invented by E. Post) is given below.

A set is said to be *immune* if it is infinite but does not contain infinite enumerable subsets. An enumerable set is said to be *simple* if its complement is immune. (Obviously, such a set cannot be decidable.)

Theorem 14. *There exists a simple set.*

Proof. We need an enumerable set S with the immune complement. This means that S must have a common point with any infinite enumerable set. To guarantee this, we will add to S an element of each enumerable set V (at least, for infinite V). In so doing, we have to ensure that infinitely many elements are left outside S. To this end, we will add to S only sufficiently large elements (we will add only one element from the ith set, and choose it to be greater than $2i$).

Let us explain the construction in more detail. Suppose that W is a universal enumerable set of pairs. Its sections W_i are all enumerable sets of natural numbers. Let the section W_i be called "the enumerable set number i" (it is possible that different numbers are assigned to the same set). Consider the set of pairs $T = \{\langle i, x \rangle \mid (x \in W_i) \text{ and } (x > 2i)\}$. This set is enumerable (as the intersection of W and the decidable set $\{\langle i, x \rangle \mid x > 2i\}$). Let us enumerate T, omitting the pairs whose first element has already occurred earlier in the enumeration. Then a certain enumerable subset T' of the set T

remains. Now consider the enumerable set S of the second elements of all pairs contained in T'.

The set S has a nonempty intersection with any infinite enumerable set. Indeed, since any infinite section W_i contains numbers greater than $2i$, the set T contains at least one pair with the first component i, and the set T' contains exactly one pair with the first component i. The second component of this pair belongs both to S and to W_i.

On the other hand, the set S has infinite complement, since at most $n+1$ different numbers from 0 to $2n$ belong to S (these are the numbers that get into S from one of the first n vertical sections; all other numbers in S are greater than $2n$). $\qquad\square$

Problem 26. Prove that an infinite set that does not contain infinite decidable subsets is immune.

Problem 27. Prove that there exists an enumerable set A with the following property: the complement of A is infinite and the increasing sequence $\alpha(0) < \alpha(1) < \alpha(2) < \cdots$ of all elements of $\mathbb{N} \setminus A$ has no computable upper bound (i.e., for any total computable function b there exists a number n such that $\alpha(n) > b(n)$). Prove that any enumerable set A with this property is simple.

Chapter 3

Numberings and Operations

1. Gödel universal functions

It is obvious that the composition of two computable functions is computable. Moreover, this statement seems to be "effective" in the sense that from the programs of two functions we can algorithmically obtain a program of their composition. In a reasonable programming language, it will consist of two procedures corresponding to two computable functions and the main program with the single line "return (f(g(x)))".

However, in order to avoid the details of programming languages, we prefer to speak about numbers of functions rather then programs. And we have tools to do so. Namely, any function U universal for the class of computable functions of one variable specifies a numbering of this class: a natural number n is a number of the function $U_n \colon x \mapsto U(n,x)$.

In general, a *numbering* (more exactly, *natural numbering*) of an arbitrary set $\mathcal{F}$ is a total map $\nu \colon \mathbb{N} \to \mathcal{F}$ whose range is the entire set $\mathcal{F}$. If $\nu(n) = f$, then n is called a *number* of the object f. Thus any binary function specifies a numbering of a certain class of unary functions (and is universal for this class).

Our current goal is to give an accurate formulation and proof for the following statement: (under certain conditions on a numbering of computable functions) there exists an algorithm that assigns to numbers of any two computable functions a number of their composition.

First of all we require that the universal function that specifies the numbering is computable. (Such numberings are called *computable*.) However, this condition alone is not sufficient: what we need is a *Gödel numbering*. Let U be a binary computable universal function for the class of unary computable functions. It is called a *Gödel universal function* if for any computable binary function V there exists a total computable unary function s such that

$$V(m, x) = U(s(m), x)$$

for all m and x (the equality is understood, as usual, in the sense that either both values are undefined or they are defined and equal to each other).

In other words, $V_m = U_{s(m)}$, that is, the function s assigns to a V-number of a function a certain U-number of the same function.

Theorem 15. *A Gödel universal function exists.*

Proof. We give two proofs of this theorem.

In the first proof we show that the construction of a universal function described in the proof of Theorem 6 (p. 12) yields a Gödel universal function. Recall that we have enumerated all the programs $p_0, p_1, p_2, \ldots$ in some natural programming language in ascending order of their lengths and defined $U(n, x)$ to be the result produced by the program p_n applied to the input x. Now suppose that V is some other computable function of two variables. For any natural m, we must obtain the program of the function V_m, that is, of the function obtained by setting the first argument of V equal to m. Clearly, in any reasonable programming language, such a program can easily be created: it suffices simply to substitute a constant for the first argument in the program of V (or transform the program of V into a procedure that is called, with a fixed first argument, from the main program).

However, it is not necessary to go into the details of the construction of a universal function; our second proof uses only the existence of a universal function.

First, we notice that there exists a ternary computable function universal for the class of binary computable functions, that is, a function T such that all binary computable functions occur among the functions $T_n(u, v) = T(n, u, v)$ obtained by fixing the first argument of T.

Such a function T can be constructed as follows. We fix an arbitrary computable numbering of pairs, that is, a computable one-to-one correspondence $\langle u, v \rangle \leftrightarrow [u, v]$ between $\mathbb{N} \times \mathbb{N}$ and $\mathbb{N}$; the integer $[u, v]$ corresponding to the pair $\langle u, v \rangle$ will be called the *number* of this pair. Let R be a binary computable universal function for the class of all unary computable functions. Then the computable function T defined by the formula $T(n, u, v) = R(n, [u, v])$ is universal for binary computable functions. Indeed, let F be an arbitrary binary computable function. Consider the unary computable function f defined by the relation $f([u, v]) = F(u, v)$. Since R is universal, there exists a number n such that $R(n, x) = f(x)$ for all x. For this n, we have the relations $T(n, u, v) = R(n, [u, v]) = f([u, v]) = F(u, v)$; hence the nth section of the function T coincides with F. This means that T is the desired ternary universal function.

Now we will use T to define a binary Gödel universal function U. Informally, we will build into U all other binary computable functions; thus U will become a Gödel function. To formalize this idea, we set $U([n, u], v) = T(n, u, v)$. Let us show that the function U thus obtained is Gödel. Any binary computable function V occurs among the sections of the function T: we can find n such that $V(u, v) = T(n, u, v)$ for all u and v. Then $V(u, v) = U([n, u], v)$ for all u and v, and hence the function s defined by $s(u) = [n, u]$ satisfies the requirement mentioned in the definition of Gödel universal function. $\qquad \square$

The numberings of computable functions that correspond to Gödel universal functions are called *Gödel numberings*.

Now we are ready to prove the precise version of the statement mentioned at the beginning of this chapter.

Theorem 16. *Let U be a binary Gödel universal function for the class of unary computable functions. Then there exists a total function c that assigns to numbers p and q of two unary functions a number $c(p, q)$ of their composition: $U_{c(p,q)}$ is the composition $U_p \circ U_q$, or*

$$U(c(p, q), x) = U(p, U(q, x))$$

for all p, q, and x.

Proof. Consider a binary computable function V defined by the equation $V([p, q], x) = U(p, U(q, x))$. By the definition of a Gödel universal function, there exists a unary total computable function s such that $V(m, x) = U(s(m), x)$ for all m and x. Then $V([p, q], x) = U(s([p, q]), x)$. Hence the function c defined by the relation $c(p, q) = s([p, q])$ is the desired one.   □

Let us repeat this proof informally. (For brevity, we identify a program and its number and consider a number m as the U-program of the function U_m.) The function V can be considered as an interpreter for some programming language. According to the definition of a Gödel numbering, there exists an algorithm s that converts any V-program m into a U-program $s(m)$ of the same function.

Now we construct a new programming language declaring that a pair $\langle p, q \rangle$ is a program of the composition of the functions with U-programs p and q. By assumption, such a program can be algorithmically converted into a U-program. This completes the proof.

It is interesting that the statement converse to Theorem 16 is also true:

Problem 28. Let U be a binary computable universal function for the class of unary computable functions. If there exists a total function that assigns to any p and q some U-number of the composition of functions that have U-numbers p and q, then U is a Gödel universal function. (*Hint*: show that a U-number of the function $x \mapsto [k, x]$ can be algorithmically obtained from k.)

A natural question arises: do there exist computable universal functions that are not Gödel? Later we will see that they do exist.

Problem 29. Let us change the definition of a Gödel universal function and require the converter s to exist only for universal computable functions V (rather than for all functions, as before). Show that the new definition is equivalent to the old one. (*Hint*: any function can be artificially reworked into a universal one by "interleaving" it with some universal function.)

Problem 30. Let U be a Gödel universal function. Prove that for any computable function $V(m, n, x)$ there exists a total computable function $s(m, n)$ such that $V(m, n, x) = U(s(m, n), x)$ for all m, n, and x. (*Hint*: combine m and n into a pair.)

2. Computable sequences of computable functions

Let $f_0, f_1, \ldots$ be a sequence of computable functions of one variable. We want to assign a meaning to the expression "a sequence $i \mapsto f_i$ is computable". There are two natural definitions:

- this sequence is called computable if the binary function F defined by the formula $F(i, n) = f_i(n)$ is computable;

- this sequence is computable if there exists a computable sequence of natural numbers $c_0, c_1, \ldots$ such that c_i is one of the numbers of the function f_i (for each i).

The second definition (unlike the first) depends on the numbering of computable functions.

Theorem 17. *If the numbering is computable (that is, the corresponding universal function U is computable), then the second definition implies the first. If, in addition, it is a Gödel numbering, then the first definition implies the second.*

(In what follows, speaking about a computable sequence of computable functions, we will always assume that the numbering is Gödel, so either definition will be good.)

Proof. If U is a computable universal function and the sequence $i \mapsto c_i$ is computable, then the function $F\colon \langle i, x \rangle \mapsto f_i(x) = U(c_i, x)$ is computable as the composition of computable functions.

Conversely, if a function F is computable and a universal function U is Gödel, then the converter that exists by the definition of a Gödel universal function is just the function that takes i into one of the numbers of the function f_i. □

Problem 31. Let us fix a Gödel universal function for the class of unary computable functions. In line with the definition on p. 9, it specifies a numbering of computable reals: a number of a computable real α is any number of any function that assigns to each rational $\varepsilon > 0$ an ε-approximation to α.

(**a**) Show that there exists an algorithm that computes one of the numbers of the sum of two computable real numbers from arbitrary numbers of the summands.

(**b**) Show that there is no algorithm that determines from any number of an arbitrary computable real x whether x is equal to zero.

(**c**) As we have seen in Problem 14, any computable real number has a computable decimal representation. Show that, nonetheless, no algorithm can transform any number of any computable real x into a number of the computable function that specifies the decimal representation of x.

3. Gödel universal sets

Let us carry over the definitions given above for computable functions to the case of enumerable sets. An enumerable set $W \subset \mathbb{N} \times \mathbb{N}$ is called a *Gödel universal enumerable set* (for the class of all enumerable subsets of $\mathbb{N}$) if for any enumerable set $V \subset \mathbb{N} \times \mathbb{N}$ there exists a total computable function $s \colon \mathbb{N} \to \mathbb{N}$ such that

$$\langle n, x \rangle \in V \Leftrightarrow \langle s(n), x \rangle \in W$$

for all n and x. (Obviously, this property implies universality.)

As in the case of functions, we can switch to numberings. Each set $U \subset \mathbb{N} \times \mathbb{N}$ defines a numbering of a certain family of subsets of $\mathbb{N}$ in the following way: n is a number of the nth section $U_n = \{x \mid \langle n, x \rangle \in U\}$. An enumerable subset of $\mathbb{N} \times \mathbb{N}$ specifies a numbering of a certain family of enumerable subsets of $\mathbb{N}$; such numberings are called *computable*. An enumerable set $W \subset \mathbb{N} \times \mathbb{N}$ is *universal* if and only if any enumerable subset of the natural numbers has a W-number; W is a *Gödel* set if and only if any computable numbering V (of any family of enumerable sets) is computably reducible to the W-numbering in the sense that $V_n = W_{s(n)}$ for some computable function s and for all n.

Theorem 18. *A Gödel universal enumerable set $W \subset \mathbb{N} \times \mathbb{N}$ exists.*

Proof. This theorem is an obvious corollary to the following statement:

Lemma. *The domain of a Gödel universal function for the class of unary computable functions is a Gödel universal set for the class of enumerable subsets of $\mathbb{N}$.*

Proof of the Lemma. Let U be a Gödel universal function, and let W be its domain. Consider an arbitrary enumerable set $V \subset \mathbb{N} \times \mathbb{N}$ and a computable function G with domain V. Since the function U is Gödel, there exists a total computable function $s \colon \mathbb{N} \to \mathbb{N}$ such that $G_n = U_{s(n)}$ for all n. Then the domains of the functions G_n and $U_{s(n)}$ also coincide, that is, $V_n = W_{s(n)}$. $\square$

Problem 32. Construct a Gödel universal set directly, using a universal subset of $\mathbb{N}^3$ (proceed by analogy with the above construction of a Gödel universal function).

Now, along the same lines as in the case of functions, we can prove that various set operations correspond to computable transformations of their numbers. Here is one example of this kind.

Theorem 19. *Let $W \subset \mathbb{N} \times \mathbb{N}$ be a Gödel universal enumerable set. Then a number of the intersection of two enumerable sets can be algorithmically computed from W-numbers of these sets: there exists a binary total computable function s such that*

$$W_{s(m,n)} = W_m \cap W_n$$

for any m and n.

Proof. Consider the set $V \subset \mathbb{N} \times \mathbb{N}$ defined by the relation

$$\langle [m,n], x \rangle \in V \Leftrightarrow x \in (W_m \cap W_n)$$

(the brackets here denote the pair's number) and apply the definition of a Gödel universal set to V. □

As in the case of functions, the notion of computability of a sequence of enumerable sets can be defined in two ways: one way is to call such a sequence computable if it coincides with the sequence $V_0, V_1, \ldots$ of sections of an enumerable set V; the other approach is to require that there is an algorithm that for any given i computes one of the numbers of the ith set in a Gödel numbering. These definitions are equivalent (the proof is similar to the one for functions).

Chapter 4

Properties of Gödel Numberings

1. Sets of numbers

We begin with an example. Consider the set of numbers of the empty function in a certain Gödel numbering. Is it decidable? In other words, is it possible to determine whether a function is empty given its number in a Gödel numbering?

Before we try to answer this question, let us notice that the answer does not depend on the choice of a Gödel numbering. Indeed, any two different Gödel numberings can be "reduced" to each other: given a number of a function in one numbering, it is possible to algorithmically obtain a number of the same function in the other numbering. If, using one numbering, we could test whether a function is empty, then, applying the "transition functions", we could just as well do this in the other numbering.

The following theorem shows that the answer to our initial question is negative.

Theorem 20. *Let U be an arbitrary Gödel universal function. Then the set of all the numbers n such that the function U_n is empty is undecidable.*

Proof. We use the method called "reduction". The idea is to show that if the set in question were decidable, then any enumerable set would be decidable as well. (And this, as we know, is not true.)

Let K be an arbitrary enumerable undecidable set. Consider the following binary computable function V:

$$V(n,x) = \begin{cases} 0 \text{ if } n \in K, \\ \text{undefined if } n \notin K. \end{cases}$$

The second argument of this function is not really used, and, essentially, V is the semicharacteristic function of the set K. Obviously, the function V has sections of two kinds: for $n \in K$, the section V_n is the zero function, and for $n \notin K$, it is the empty function.

Since U is a Gödel universal function, there exists a total computable function s such that $V(n,x) = U(s(n),x)$ for all n and x, that is, $V_n = U_{s(n)}$. So for $n \in K$, the value $s(n)$ is a U-number of the zero function, and for $n \notin K$ the value $s(n)$ is a U-number of the empty function. Therefore, if the set of U-numbers of the empty function were decidable by a certain algorithm, we could apply this algorithm to $s(n)$ and find out whether the number n belongs to K or not. This would mean that K is a decidable set, contrary to our assumption.   □

In particular, we conclude that the empty function has infinitely many numbers in any Gödel numbering (because any finite set is decidable).

Furthermore, the set of numbers of the empty function is not only undecidable, it is not enumerable as well. Indeed, its complement, the set of numbers of all functions with a nonempty domain, is enumerable (This is true for any computable, not only Gödel, numbering: we can compute $U(n,x)$ in parallel for all n and x and print n whenever $U(n,x)$ is defined for at least one x.) But by Post's Theorem (p. 7), if the complement of an undecidable set is enumerable, then the set itself is nonenumerable.

Now let us prove a more general statement, sometimes called the Rice–Uspensky Theorem. Denote by $\mathcal{F}$ the class of all unary computable functions.

Theorem 21. *Let $\mathcal{A} \subset \mathcal{F}$ be an arbitrary nontrivial property of computable functions ("nontrivial" means that there are both functions satisfying the property and functions not satisfying it; that is, the set $\mathcal{A}$ is nonempty and does not coincide with $\mathcal{F}$). Let U be a Gödel universal function. Then it is impossible to determine algorithmically whether a computable function with a given U-number has the property $\mathcal{A}$. In other words, the set $\{n \mid U_n \in \mathcal{A}\}$ is undecidable.*

Proof. Without loss of generality we may assume that the empty function (we denote it by ζ) belongs to $\mathcal{A}$. Let ξ be an arbitrary function in $\mathcal{F} \setminus \mathcal{A}$. (If $\zeta \notin \mathcal{A}$, the set $\mathcal{A}$ can be replaced by its complement.)

Now we repeat the proof of the previous theorem, but replace the zero function in this proof by the function ξ: for any enumerable undecidable K, we set

$$V(n, x) = \begin{cases} \xi(x) \text{ if } n \in K, \\ \text{undefined if } n \notin K. \end{cases}$$

Again, the function V is computable (for any given n and x, we enumerate the set K waiting for n to appear and then compute $\xi(x)$). For $n \in K$, the function V_n coincides with ξ; for $n \notin K$, it coincides with ζ. Therefore, $V_n \in \mathcal{A}$ if and only if $n \notin K$. Suppose that the statement of the theorem is false. Then, by the same argument as in the previous proof, we conclude that the property $V_n \in \mathcal{A}$ is algorithmically decidable (given n). Hence we can find out whether the number n belongs to the set K or not, which is impossible by the choice of K. $\qquad\square$

A certain shortcoming of this proof is its asymmetry (the functions chosen inside and outside $\mathcal{A}$ are of a different nature: one of them is empty, the other one is arbitrary). A more symmetric version follows.

Let us show that if it were possible to recognize the property $\mathcal{A}$ by U-numbers, then any two disjoint enumerable sets P and Q could be separated by a decidable set. Choose any two functions ξ and η

"on different sides" of $\mathcal{A}$. Consider the function

$$V(n,x) = \begin{cases} \xi(x) \text{ if } n \in P, \\ \eta(x) \text{ if } n \in Q, \\ \text{undefined if } n \notin P \cup Q. \end{cases}$$

This function is computable: for any given n and x we wait until n appears either in P or in Q; then we compute $\xi(x)$ or $\eta(x)$, respectively.

If $n \in P$, then V_n coincides with ξ; if $n \in Q$, then V_n coincides with η. Therefore, by verifying whether V_n belongs to the class $\mathcal{A}$, we would be able to decidably separate P from Q. But this contradicts Theorem 13 and thus completes this more symmetric version of the proof.

The second proof shows that a stronger version of the theorem holds: for any two distinct computable functions φ and ψ and any Gödel universal function U, the set of all U-numbers of the function φ cannot be separated from the set of U-numbers of ψ by a decidable set. (Later we will see that these two sets are nonenumerable.)

Now it is easy to construct an example of a computable universal function that is not Gödel. It will suffice to ensure that the empty function has only one number. This is not difficult. Let $U(n,x)$ be an arbitrary computable universal function. Consider the set D of all U-numbers of all functions with nonempty domain. As we have already said, this set is enumerable. Consider a total computable function d that enumerates it: $D = \{d(0), d(1), \dots\}$. Now consider the function $V(i,x)$ such that $V(0,x)$ is undefined for any x and $V(i+1,x) = U(d(i),x)$. In other words, the function V_0 is empty, and the function V_{i+1} coincides with $U_{d(i)}$. It is easy to see that the function V is computable; by construction, it is universal, and the only V-number of the empty function is 0.

In fact, there exist even more exotic numberings: as Friedberg showed, one can construct a universal computable function such that any computable function has exactly one number. Obviously, such a numbering cannot be Gödel. Here is a curious rewording of Friedberg's theorem: it is possible to invent a programming language such that each programming problem has a unique solution in it. The proof of this theorem is difficult and we omit it; see Friedberg's original article (Journal of Symbolic Logic **23** (1958), 309–318) or A. I. Mal'tsev's book "Algorithms and Recursive Functions" [**6**].

A similar statement is true for numberings of enumerable sets.

2. New numbers of old functions

The Rice–Uspensky Theorem shows that the set of numbers of any specific function in a Gödel numbering is undecidable, and hence infinite. Now we will prove a stronger fact: given a number of any function in a Gödel numbering, one can algorithmically obtain arbitrary many other numbers of the same function. Formally, this can be stated as follows:

Theorem 22. *Let U be a Gödel universal function. Then there exists a total binary function g such that for any i the values $g(i,0), g(i,1), \ldots$ are different U-numbers of the function U_i.*

Proof. To prove this theorem, we construct another programming language such that any given function U_i has infinitely many programs. Then, using the fact that U is a Gödel function, we convert these programs and obtain U-numbers of the function in question. Of course, we need special precautions to ensure that we get infinitely many different U-numbers. (For example, if our programming language allows comments, then it is easy to construct infinitely many programs for the same function. However, this does not help us much, since the converter to U-numbers may start its work by deleting comments. So we need a deeper idea to succeed.)

We proceed as follows. Let h be an arbitrary function. We show that there exists an algorithm that finds infinitely many different U-numbers of the function h. (The theorem says that this can be done not only for a single specific function h, but also for all the functions U_i "uniformly in i"; for the time being we ignore this problem.)

Let P be an enumerable undecidable set. Consider a computable function

$$V(n,x) = \begin{cases} h(x) & \text{if } n \in P, \\ \text{undefined} & \text{if } n \notin P. \end{cases}$$

There are only two distinct functions among V_n: if $n \in P$, then $V_n = h$; if $n \notin P$, then V_n is the empty function ζ. We start with the case $h \neq \zeta$ (the case $h = \zeta$ needs a more sophisticated construction and is considered below).

Since U is a Gödel universal function, there exists a converter s that transforms V-numbers into U-numbers. In other terms,

- $n \in P \Rightarrow U_{s(n)} = V_n = h;$
- $n \notin P \Rightarrow U_{s(n)} = V_n = \zeta.$

It follows that if $p(0), p(1), \ldots$ is a computable enumeration of the set P, then all the numbers $s(p(0)), s(p(1)), \ldots$ are U-numbers of the function h. Let us show that the set $\{s(p(0)), s(p(1)), \ldots\}$ is infinite (and therefore, we can compute $s(p(0)), s(p(1)), \ldots$ until a new, yet unused number of the function h appears).

Suppose that this is not the case, and the set $X = \{s(n) \mid n \in P\}$ is finite. Then X is decidable. If $n \in P$, then, by construction, $s(n) \in X$; if $n \notin P$, then $s(n)$ is a number of the function ζ and does not belong to X (recall that $h \neq \zeta$ by our assumption). Hence $n \in P$ if and only if $s(n) \in X$, and so the decidability of X implies the decidability of P, contrary to our assumption.

However, this reasoning does not work if $h = \zeta$. Although $s(p(0)), s(p(1)), \ldots$ in this case are numbers of h, there is nothing to guarantee that infinitely many of them are pairwise distinct. So in this case we will use a more delicate argument and consider any computable function ξ distinct from ζ (for instance, we can set $\xi(n) = 0$ for all n). Consider two enumerable inseparable sets P and Q and the computable function

$$
V(n, x) = \begin{cases} h(x) \text{ if } n \in P, \\ \xi(x) \text{ if } n \in Q, \\ \text{undefined if } n \notin P \cup Q. \end{cases}
$$

Let s be a converter mapping V-numbers to U-numbers. Then

- $n \in P \Rightarrow U_{s(n)} = h;$
- $n \in Q \Rightarrow U_{s(n)} = \xi;$
- $n \notin P \cup Q \Rightarrow U_{s(n)} = \zeta.$

As before, $s(p(0)), s(p(1)), \ldots$ are numbers of the function h. Let us show that if $h \neq \xi$, then the set X of the numbers $s(p(0)), s(p(1)), \ldots$ is undecidable (and hence infinite). Indeed, if X were decidable, then we could separate P from Q by a decidable set. Namely, the set $\{n \mid s(n) \in X\}$ contains P (by construction) and is disjoint with Q (since for $n \in Q$, $s(n)$ is a number of the function ξ; therefore, $s(n)$ cannot belong to X).

Thus we have two constructions that allow us to obtain other numbers of a given function h. The first of them definitely succeeds (i.e., produces infinitely many new numbers) if $h \neq \zeta$; the second works if $h = \zeta$. However, we do not know in advance whether $h = \zeta$ or not. What can we do? Let us apply both constructions in parallel until one of them yields the desired new number; we know that none of them will ever produce an invalid result (i.e., numbers of another function) and at least one of them will succeed, although we do not know which of the two.

This enables us to generate new numbers of the ith function for any given i uniformly in i. Formally, we should consider the two computable binary functions V_1 and V_2 defined by the relations

$$V_1([i,n], x) = \begin{cases} U(i,x) \text{ if } n \in P, \\ \text{undefined if } n \notin P, \end{cases}$$

$$V_2([i,n], x) = \begin{cases} U(i,x) \text{ if } n \in P, \\ 0 \text{ if } n \in Q, \\ \text{undefined if } n \notin P \cup Q, \end{cases}$$

(here P and Q are fixed enumerable inseparable sets, $[u,v]$ is the number of the pair $\langle u, v \rangle$ in a fixed computable numbering of pairs). Since U is a Gödel universal function, we can find computable total functions s_1 and s_2 such that $V_1([i,n], x) = U(s_1([i,n]), x)$ and $V_2([i,n], x) = U(s_2([i,n]), x)$. Let p be a total unary function such that $P = \{p(0), p(1), \ldots\}$. Then the desired function g can be defined as follows: $g(i,k)$ is the kth number (not counting repetitions) in the sequence

$$s_1([i, p(0)]), s_2([i, p(0)]), s_1([i, p(1)]), s_2([i, p(1)]),$$
$$s_1([i, p(2)]), s_2([i, p(2)]), \ldots.$$

$\square$

3. Isomorphism of Gödel numberings

The statement about the generation of new numbers proved just above will play a crucial role in the proof of Rogers' Theorem that says that any two Gödel numberings are isomorphic. Here is the exact statement of this theorem.

Theorem 23. *Let U_1 and U_2 be two Gödel universal functions for the class of unary computable functions. Then there exist two total mutually inverse computable functions s_{12} and s_{21} such that*

$$U_1(n, x) = U_2(s_{12}(n), x) \quad and \quad U_2(n, x) = U_1(s_{21}(n), x)$$

for any n and x.

This theorem shows that we can choose converters reducing one Gödel numbering to another and vice versa to be mutually inverse, and so any two Gödel numberings differ from each other only by a computable permutation of numbers.

Notice that Theorem 22 follows from Rogers' Isomorphism Theorem. Indeed, for some Gödel numberings, e.g., for the numberings derived from typical programming languages, it is easy to obtain arbitrarily many programs of the same function (by adding comments, null statements, dummy variables, etc.); therefore, by the isomorphism, this is possible for any other Gödel numbering as well.

Proof. We will proceed along the same lines as in the proof of the isomorphism of countable dense well-ordered sets without the first and last elements. The desired bijections are constructed step by step. At the kth step a certain one-to-one correspondence

$$a_1 \leftrightarrow b_1, a_2 \leftrightarrow b_2, \ldots, a_k \leftrightarrow b_k$$

between two finite k-element subsets of the set of natural numbers is constructed. For each i, the numbers a_i and b_i in this construction are the numbers of the same function in different numberings (the a_ith function in U_1-numbering equals the b_ith function in U_2-numbering).

At each step of the construction we add a new pair $a_k \leftrightarrow b_k$ preserving the above-mentioned property. In so doing, we will gradually incorporate all natural numbers on both sides. Thus we will obtain the desired one-to-one correspondence; it will be computable, because our construction is computable.

So, how do we add a new pair? We alternate steps of two types. At even steps we take the smallest natural number u that has not yet appeared on the left side of the correspondence, among the numbers a_i. This number u is a U_1-number of a certain function. Since U_2 is a Gödel numbering, we can find a U_2-number of the same function. Denote it by v. If v does not occur among the numbers b_i, we are done: we add the pair $u \leftrightarrow v$ to our correspondence. If v has already appeared in the construction, we use Theorem 22 and obtain other U_2-numbers of the same function until a new number (distinct from any of the b_i) appears.

Odd steps are similar except that we begin with the smallest number that has not yet appeared among the b_i. □

A remark for programmers (not to be taken seriously): since, say, Pascal and C can be viewed as Gödel numberings, by the theorem proved above, there exist not just some converters from Pascal to C and back, but converters that are mutually inverse. (The main part of the proof, a reference to Theorem 22, is not really used here, though: both in Pascal and C one can add comments to a program, thus creating as many equivalent programs as needed.)

A similar theorem is true for Gödel numberings of enumerable sets: any two such numberings are isomorphic (differ by a computable permutation of numbers). The proof is also similar; first we must show that from a number of a set in a Gödel numbering, arbitrarily many numbers of the same set can be algorithmically obtained. This is done using two methods: one of them works for any nonempty set; the other is applicable to the empty set.

Problem 33. Complete this argument.

4. Enumerable properties of functions

We have already seen (Theorem 21) that all (nontrivial) properties of functions are undecidable. But some of them are enumerable (in the sense that the set of all numbers of all functions with such a property is enumerable). For instance, such is the property of a function to have nonempty domain (see above). Another example: the property "$f(200)$ is defined and equals 2".

All enumerable properties can be described in a fairly simple way. This description is based on a number of definitions.

A function with natural arguments and values is called a *pattern* if its domain is finite. In other words, a pattern is a finite list of pairs ⟨argument, value⟩ where all arguments are distinct.

Patterns can be regarded as constructive objects (they can be encoded by binary strings, natural numbers, etc.). This allows us to speak about a decidable set of patterns, an enumerable set of patterns, and so on.

For each pattern t, consider the property of a function "to be an *extension* of t", i.e., the set $\Gamma(t)$ of all (computable) functions that are extensions of t. [Note that the sets $\Gamma(t)$ form a base of topology on the set of all computable functions.] It is easy to see that for any t and for any computable numbering U, the set of all numbers of all functions from $\Gamma(t)$ is enumerable. Indeed, we can compute the values $U_n(x)$ for all n and x in parallel; as soon as the cumulative data allows us to assert that $U_n \in \Gamma(t)$, the number n is printed. (If $U_n \in \Gamma(t)$, then this fact will be discovered after a finite number of steps, because the domain of the pattern t is finite.)

Let T be an arbitrary set of patterns. Denote by $\Gamma(T)$ the set of all computable functions that extend at least one pattern from T, i.e., the union of sets $\Gamma(t)$ over all $t \in T$. Now, as we promised above, we can give a description of enumerable properties.

Theorem 24. (a) *Let T be an arbitrary enumerable set of patterns, and let U be a computable universal function for the class of all unary computable functions. Then the set of all U-numbers of all functions from $\Gamma(T)$ is enumerable.* (b) *Let U be a Gödel universal function (for the class of all unary computable functions). Let $\mathcal{G}$ be a subset* *of this class. If the set $\{n \mid U_n \in \mathcal{G}\}$ of all U-numbers of all functions from the class $\mathcal{G}$ is enumerable, then $\mathcal{G} = \Gamma(T)$ for some enumerable set T of patterns.*

For instance, the class of functions with a nonempty domain (mentioned above) corresponds to the set of all nonempty patterns (as well as, e.g., to the set of all patterns with a one-element domain). And the property "$f(200) = 2$" corresponds to the pattern $\langle 200, 2 \rangle$ that consists of one pair.

Proof. Statement (a) is easy: we compute all values of $U(n, x)$ and enumerate all patterns from T in parallel; as soon as we find out that one of the functions U_n is an extension of a pattern from T, we print n.

More substantial is statement (b), in which we use the fact that the numbering in question is Gödel. We will need two lemmas.

Lemma 1. *If a computable function h is an extension of a computable function g of the class $\mathcal{G}$, then the function h also belongs to the class $\mathcal{G}$.*

Lemma 2. *If a computable function g belongs to the class $\mathcal{G}$, then g is an extension of some pattern h in $\mathcal{G}$, i.e., of a function $h \in \mathcal{G}$ with a finite domain.*

[These two lemmas together mean that any enumerable property $\mathcal{G}$ is open in the topology described above.]

Let us show how statement (b) of our theorem is derived from these lemmas. Notice that the set T of all patterns from the class $\mathcal{G}$ is enumerable. Indeed, since U is a Gödel function, from a pattern as a constructive object (i.e., from the list of pairs $\langle \text{argument}, \text{value} \rangle$), we can compute its U-number. (Formally: we consider the function $\langle t, x \rangle \mapsto$ (value of the pattern t at point x) and apply the definition of Gödel universal functions.) Therefore, the set T is the preimage of the enumerable set of all U-numbers of all functions from the class $\mathcal{G}$; hence T itself is enumerable.

Lemmas 1 and 2 ensure that $\mathcal{G} = \Gamma(T)$. Indeed, by Lemma 1, any function from $\Gamma(T)$ belongs to the class $\mathcal{G}$, since it is an extension of a pattern from $\mathcal{G}$. On the other hand, Lemma 2 ensures that any function g from $\mathcal{G}$ is an extension of a pattern from $\mathcal{G}$ (and thus from T); therefore, $g \in \Gamma(T)$.

It remains to prove Lemmas 1 and 2. Suppose that, contrary to Lemma 1, there is a function g that belongs to the class $\mathcal{G}$ and has an extension h not in this class. Let us take an enumerable undecidable set K and consider the following function of two arguments:

$$V(n, x) = \begin{cases} h(x) \text{ if } n \in K, \\ g(x) \text{ if } n \notin K. \end{cases}$$

This function is computable. Indeed, it can be readily seen that its graph is enumerable as the union of the graph of g multiplied by $\mathbb{N}$ and the graph of h multiplied by K. In other words, to compute $V(n, x)$, we start the process of enumeration of K and (in parallel with this process) the computation of $g(x)$ and $h(x)$. A result is returned if the computation of $g(x)$ has been completed (in this case it does not matter whether n lies in K, since h is an extension of g) or if the computation of $h(x)$ has been completed and, in addition, n has been detected in K.

Since U is a Gödel function, there exists a total function s with the following property:

- $n \in K \Rightarrow U_{s(n)} = h \Rightarrow U_{s(n)} \notin \mathcal{G}$;
- $n \notin K \Rightarrow U_{s(n)} = g \Rightarrow U_{s(n)} \in \mathcal{G}$.

Thus the complement of K is the preimage of the enumerable set of all numbers of all functions from $\mathcal{G}$ under the computable mapping s. Hence, by Theorem 5, this complement is enumerable. This is a contradiction that completes the proof of Lemma 1.

Lemma 2 is proved similarly. Suppose that (contrary to the statement of the lemma) the function g belongs to the class $\mathcal{G}$, but none of its finite parts belongs to this class. Consider the function

$$V(n, x) = \begin{cases} g(x) & \text{if after } x \text{ steps of the enumeration of } K \\ & n \text{ has not yet appeared,} \\ \text{undefined if } n \text{ has appeared.} \end{cases}$$

It can be readily seen that for $n \notin K$, the function V_n coincides with g and hence belongs to $\mathcal{G}$, whereas for $n \in K$, the function V_n is a finite part of the function g and hence does not belong to $\mathcal{G}$. The proof is completed as in Lemma 1. □

Chapter 5

Fixed Point Theorem

1. Fixed point and equivalence relations

Theorem 25. *Let U be a Gödel universal computable function for the class of unary computable functions, and let h be an arbitrary unary total computable function. Then there exists a number n such that $U_n = U_{h(n)}$, that is, n and $h(n)$ are numbers of the same function.*

In other words, it is impossible to find an algorithmic transformation of programs that would assign to each program another, nonequivalent one. This theorem is called *Kleene's Fixed Point Theorem* or the *Recursion Theorem*.

Proof. We will use the construction of a computable function that has no total computable extension (Chapter 2).

Consider an arbitrary equivalence relation (which will be denoted by $x \equiv y$) on the set of natural numbers. We will show that the following two properties of this relation cannot hold simultaneously:

- For any computable function f, there exists a total computable function g which is its $\equiv$-extension (this means that if $f(x)$ is defined for some x, then $g(x) \equiv f(x)$).

- There exists a total computable function h that has no $\equiv$-fixed point (i.e., $n \not\equiv h(n)$ for all n).

Both Theorem 10 and the Fixed Point Theorem are corollaries of this statement. Indeed, if $x \equiv y$ is the equality relation $(x = y)$, then the second property holds (put, for instance, $h(n) = n+1$); therefore, the first one does not hold, and we conclude that there exists a computable function f that has no computable $=$-extension. To obtain the Fixed Point Theorem, we interpret $x \equiv y$ as the relation $U_x = U_y$ (x and y are numbers of the same function). In this case, as we are going to show, the first property holds; hence the second one does not.

Why does the first property hold? Let f be an arbitrary (partial) computable function of one argument. Consider the function $V(n,x) = U(f(n),x)$. Since U is a Gödel universal function, one can find a total function s such that $V(n,x) = U(s(n),x)$ for all n and x. This function is just the desired $\equiv$-extension. Indeed, if $f(n)$ is defined, then $s(n)$ is another U-number of the function with the U-number $f(n)$. (Notice that if $f(n)$ is not defined, then $s(n)$ is one of the numbers of the completely undefined function.)

To complete the proof of the Fixed Point Theorem, it remains to verify that the two properties of equivalence specified above are incompatible. This is done by means of Theorem 9 (Section 2.2). Let us take a computable function f such that no computable function differs from it everywhere (for instance, the diagonal function $x \mapsto U(x,x)$ for any computable universal function U). Suppose that there exists a total computable $\equiv$-extension g of the function f and, contrary to our alternative, a total computable function h without $\equiv$-fixed points. Consider the function $t(x) = h(g(x))$. Then t differs from f everywhere. Indeed, if $f(x)$ is defined, then $f(x) \equiv g(x) \not\equiv h(g(x)) = t(x)$ and, therefore, $f(x) \neq t(x)$. And if $f(x)$ is not defined, then it is this fact by itself that distinguishes $f(x)$ from $t(x)$. This contradicts the choice of the function f and thus completes the proof. $\square$

The Fixed Point Theorem can be restated as follows:

Theorem 26. *Let $U(n,x)$ be a Gödel computable universal function for the class of unary computable functions. Let $V(n,x)$ be an arbitrary computable function. Then the functions U and V coincide* *on a certain section: there exists a number p such that $U_p = V_p$, or $U(p, n) = V(p, n)$ for any n.*

Proof. Since U is a Gödel function, we can find a total computable function h such that $V(n, x) = U(h(n), x)$ for all n and x. It remains to take as p the fixed point of the function h. $\square$

(Here is an example of a consequence of this theorem: no matter how inventive programmers are, for any two versions of a compiler they can ever work out, there will be a program that behaves the same way in both versions, for instance, in both cases it will get into a loop. The only chance to create "completely incompatible versions" (such that no program behaves the same way in both versions) is to construct a compiler that is not a Gödel universal function. In fact, our programmers may succeed, but only if the function specified by their compiler is not Gödel universal. However, one must try really hard for this to happen!)

It is instructive to trace the construction of the fixed point in more detail. To make it easier to grasp, let us write $[n](x)$ instead of $U(n, x)$, and read this notation as "the result of application of the program n to the input x."

The argument starts with the consideration of the "diagonal" function $U(x, x)$, which can now be written as $[x](x)$ (the result of application of the program x to itself). Then we construct its total $\equiv$-extension, which is done as follows. The expression $[[x](x)](y)$ computably depends on two arguments. We recall that U is a Gödel universal function and find a program g applicable to any x such that $[[g](x)](y) = [[x](x)](y)$ for all x and y. We are looking for a fixed point of the program h. We consider the composition $[h]([g](x))$; it depends on x computably. Therefore, there exists a program t such that $[t](x) = [h]([g](x))$ for all x. This program is applicable to all x, because so are h and g. Now $[g](t)$ is the fixed point. Indeed, we must check that $[[g](t)](x) = [[h]([g](t))](x)$ for all x. But by the definition of g, we have $[[g](t)](x) = [[t](t)](x)$, and recalling the definition of t, we can rewrite the right-hand side of this relation as $[[h]([g](t))](x)$, which is just what we need.

2. A program that prints its text

The following corollary is a classical application of the Fixed Point Theorem: there exists a program that prints (on any input) its own text. Indeed, if such a program did not exist, then the map

$$p \mapsto \text{(the program that prints } p \text{ on any input)}$$

would have no fixed point.

Formally, this corollary can be expressed as follows:

Theorem 27. *Let $U(n,x)$ be a Gödel universal computable function for the class of all unary computable functions. Then there exists a number p such that $U(p,x) = p$ for any x.*

Let us say this in programmers' language. Let $U(p,x)$ be the result of application of a Pascal program p to the standard input x. (*Refinements*: (1) we identify numbers and finite sequences of bytes; (2) if a program never terminates, we assume that its result is undefined, even if there is something sent to the standard output.) Clearly, the function U is a Gödel universal function. Applying the above theorem to this function, we see that there is a program p that returns p at the output whatever the input is.

The Pascal programming language mentioned above is as good as any other: the argument will clearly remain true anyway.

Problem 34. Prove that there exists a Pascal program that prints its text back to front.

Problem 35. Show that there are two different Pascal programs P and Q such that the program P prints the text of the program Q, and Q prints the text of P. (If we do not require that P and Q be different, then we can take as P and Q the same program printing its text.)

Let us explicitly write a Pascal program that prints its text. (This is a good exercise for programming fans.) To begin with, let us write an informal instruction in English:

|| *print twice, the second time in quotes, the following text: "print twice, the second time in quotes, the following text:"*

To write something like that in Pascal, we will need some additional tricks, but the idea is clear: a string constant is used twice. Here is a version of such a program:

```
program selfprint;
var a:array[1..100]of string;i:integer;
begin
a[1]:='program selfprint;';
a[2]:=' var a:array[1..100]of string;i:integer;';
a[3]:='begin';
a[4]:='for i:=1 to 3 do writeln(a[i]);';
a[5]:='for i:=1 to 11 do begin';
a[6]:='  write(chr(97),chr(91),i);';
a[7]:='  write(chr(93),chr(58),chr(61));';
a[8]:='  writeln(chr(39),a[i],chr(39),chr(59));';
a[9]:='end;';
a[10]:='for i:=4 to 11 do writeln(a[i]);';
a[11]:='end.';
for i:=1 to 3 do writeln(a[i]);
for i:=1 to 11 do begin
  write(chr(97),chr(91),i);
  write(chr(93),chr(58),chr(61));
  writeln(chr(39),a[i],chr(39),chr(59));
end;
for i:=4 to 11 do writeln(a[i]);
end.
```

Reading this program, it is helpful to keep in mind the following relationship between symbols and their codes:

a	[	]	:	=	'	;
97	91	93	58	61	39	59

We see that this program can be easily modified to print, say, its text back to front: to this end, the commands **write** and **writeln** that print the text must be replaced by the commands that write it to a file (or an array of bytes), and then the commands that print this file or array in reverse order.

By adding another step, we can obtain a proof of the Fixed Point
Theorem. Suppose that h is a transformation of Pascal programs, and
we want to find its fixed point. Then we write a program similar to
the one above that will write its text in a string p, then apply h to p
to obtain another string q, and then launch a Pascal interpreter on
the string q ("redirecting standard input to the input of program q").
Of course, this program will be rather long, because it will include the
Pascal interpreter written in Pascal (and even twice: the first time
plainly and the second time in quotes).

Clearly, this program is a fixed point of the transformation h:
its execution begins with the computation of the value of h on the
program's text; after that, this value is understood as a program and
is applied to the input.

As a matter of fact, this proof is a direct translation of the previous one (the original proof was a "mathematician's version", while
the last one is a "programmer's version" of the same proof).

3. System trick: Another proof

If experts in various programming languages were asked to compete in
writing the shortest program that prints its text, then, most probably,
the following tiny program in BASIC would win:

```
10 LIST
```

The point is that BASIC has a command, LIST, which prints the
program text and can be launched from inside the program.

First of all, this is a good joke. But one can take this joke seriously
and use this idea in yet another proof of the Fixed Point Theorem
(to be more exact, in another version of the same proof).

To begin with, we notice that it suffices to prove the theorem for
a specific Gödel numbering of our choice. Indeed, suppose that some
other Gödel numbering admits a function without a fixed point, that
is, there is a method of converting each program into a nonequivalent
one. Then, by translating back and forth, a similar method could also
be found for the first numbering (for which the theorem is assumed
to have been proved).

Now consider a programming language which, in addition to standard constructions, has a built-in procedure:

```
GetProgramText (var s: string)
```

This procedure places the text of the original program into the string s. Although this idea is somewhat unusual, it is quite possible to imagine an interpreter of this language, and the interpretation of this procedure is not difficult, because the program text is certainly available to the interpreter. Let us make another step and imagine that the language also includes the procedure:

```
ExecuteProgram(s: string)
```

This procedure transfers control to the program whose text is in the string s assuming that this program gets the input of the original program. It is also clear what the language interpreter should do in this case: it must recursively call itself on the content of the string s and the input data.

Our enhanced programming language certainly admits translation into customary languages (since it has an interpreter) and back (since it is not necessary to use the new constructions). Therefore, the numbering of computable functions it specifies is a Gödel numbering. Let h be a total computable function whose fixed point we want to find. Consider a program in our language that computes the function h:

```
function Compute_h (x: string) : string;
begin
   ...
end;
```

(Here we do not even need language extensions.) Now let us write the program FP which is a fixed point of the function h:

```
program FP;
   var s: string;
   function Compute_h (x:string) : string;
   begin
      ...
   end;
```

```
begin
   GetProgramText (s);
   s := Compute_h (s);
   ExecuteProgram (s);
end.
```

The execution of the program FP immediately boils down to the execution of the program obtained by the application of the function h to FP, so it is a fixed point by construction.

Problem 36. Let h be the identity function, that is, $h(x) = x$. (Then, of course, any program will be its fixed point.) What is the program produced by the construction described just above? (Answer: the program that loops on any input.)

We have explained how to prove the Fixed Point Theorem using a language with the additional procedure "get the program text". But we can also reverse our reasoning and explain why the application of the Fixed Point Theorem replaces this additional procedure.

Suppose we have a program p that contains a call GetProgramText(s). Let us replace this line by the assignment operator s := t, where t is a string constant. We get a new program depending on t. Let us call it $p(t)$. By the Fixed Point Theorem, there exists a value of t for which the programs t and $p(t)$ are equivalent. For this t, the execution of the program t is equivalent to the execution of its text with the text of the program t substituted into the string s when the procedure GetProgramText(s) is called.

Now it becomes clearer why the Fixed Point Theorem is also called the Recursion Theorem. Indeed, recursion consists in calling a program from itself. But there is an important difference between our construction and ordinary recursive calls: we have the right not only to call the program, but even to get access to its text! An ordinary call is actually a particular case of access to the text, since we can call the interpretation procedure on this text. (Of course, in so doing, we will have to include in the program the text of the interpreter of our programming language written in this language.)

4. Several remarks

Infinite set of fixed points. Theorem 25 (the Fixed Point Theorem) establishes the existence of at least one fixed point. In fact, it is easy to understand that the set of fixed points is infinite: in the notation of this theorem, there exist infinitely many numbers n such that $U_n = U_{h(n)}$.

One possible explanation is this: if there were finitely many fixed points, then we would be able to change the function h at these points and to clear away all fixed points. This argument, however, does not allow us to generate fixed points effectively (to specify the infinite enumerable set of fixed points of a given function h). This can also be done. Recall the proof of Theorem 25. The fixed points in this proof turned out to be the values $[g](t)$; but the function g can be chosen so that all its values are greater than any number specified in advance (Theorem 22, p. 31).

Problem 37. Complete this argument.

A parametric version of the Fixed Point Theorem. If a program converter computably depends on a parameter, then we can choose a fixed point computably dependent on this parameter as well. The exact meaning of this statement is clarified by the following theorem.

Theorem 28. *Let U be a Gödel universal function for the class of unary computable functions, and let h be a binary total computable function. Then there exists a unary total computable function n that assigns to each p a fixed point of the function h_p, so that $U_{h(p,n(p))} = U_{n(p)}$ or, in other words,*

$$U(h(p, n(p)), x) = U(n(p), x)$$

for all p and x (as usual, both sides can be simultaneously undefined).

Proof. We have seen that a fixed point is found constructively. Therefore, if we apply our construction to get a fixed point of the function h_p (that computably depends on the parameter p), the result of our construction will also computably depend on p.

We omit the formal details; they are rather straightforward and hardly make the proof clearer. □

In this theorem we assumed that the family of functions h_p consists of total functions. As a matter of fact, this is not necessary, because for an arbitrary computable family of computable functions h_p (in other words, for an arbitrary binary computable function h), there exists a unary total computable function n with the following property: for each p either the function h_p is undefined at the point $n(p)$ or $n(p)$ is a fixed point of the function h_p.

Problem 38. Check that the function $n(p)$ constructed in the proof of Theorem 28 has this property. (One can also use $\equiv$-extension of h where $p \equiv q$ stands for $U_p = U_q$.)

Problem 39. Combining the remarks we made above, show that for any computable function h (specified by its number with respect to a fixed Gödel universal function), one can effectively find infinitely many natural numbers each of which is either a fixed point of the function h or a point at which this function is undefined.

Fixed point for enumerable sets. Similar statements are true for Gödel numberings of enumerable sets. For example, let us prove that if W is a Gödel universal enumerable set, then any computable total function h has a fixed point n such that $W_n = W_{h(n)}$.

Indeed, if W is a Gödel universal enumerable set, then the argument from the proof of Theorem 25 can be applied to the equivalence relation

$$a \equiv b \Leftrightarrow W_a = W_b,$$

because any computable function f has a total computable $\equiv$-oxtonsion.

Let us verify this. To this end, consider the set

$$V = \{\langle p, x \rangle \mid f(p) \text{ is defined and } \langle f(p), x \rangle \in W\}.$$

It is readily seen that this set is enumerable (e.g., it is the domain of the computable function $\langle p, x \rangle \mapsto w(f(p), x)$, where w is a computable function with domain W). We have $V_p = W_{f(p)}$ if $f(p)$ is defined, and $V_p = \varnothing$ if $f(p)$ is undefined. Recalling that W is a Gödel universal set, we find a total function s such that $V_p = W_{s(p)}$. Thus $W_{s(p)} = W_{f(p)}$ whenever $f(p)$ is defined, completing the proof.

Problem 40. Let W be a Gödel universal set (for the class of all enumerable subsets of natural numbers). **(a)** Show that there exists a number x for which $W_x = \{x\}$. **(b)** Show that there exist distinct numbers x and y for which $W_x = \{y\}$ and $W_y = \{x\}$.

Application. The simplest (though not very typical) application of the Fixed Point Theorem is another proof of Theorem 21 about the undecidability of properties of computable functions. Indeed, let $\mathcal{A}$ be a nontrivial property of computable functions that can be recognized by numbers of functions in a Gödel numbering U. Suppose that the function U_p satisfies this property and U_q does not satisfy it. Then the function

$$h(x) = \begin{cases} q \text{ if the function } U_x \text{ has the property } \mathcal{A}, \\ p \text{ if the function } U_x \text{ does not have the property } \mathcal{A} \end{cases}$$

is computable and has no fixed points.

Isomorphism of universal sets. Let U_1 and U_2 be two sets of pairs of natural numbers. They are called *computably isomorphic* if one can find a computable permutation (bijection) $i \colon \mathbb{N} \to \mathbb{N}$ with the following property:

$$\langle x, y \rangle \in U_1 \Leftrightarrow \langle i(x), i(y) \rangle \in U_2.$$

Theorem 29. *Any two Gödel universal sets for the class of enumerable subsets of natural numbers are computably isomorphic.*

Proof. To begin with, let us explain the difference between this theorem and the theorem about the isomorphism of Gödel numberings of enumerable sets (see the remark at the end of Section 4.3, p. 35). In that theorem, we applied an enumerable permutation only to numbers of sets, not to elements of sets. In our current notation the theorem about the isomorphism of Gödel numberings can be written as follows:

$$\langle x, y \rangle \in U_1 \Leftrightarrow \langle i(x), y \rangle \in U_2.$$

Notice that a computable permutation with respect to the second argument preserves universality: if $U \subset \mathbb{N}^2$ is a Gödel universal set and

$i\colon \mathbb{N} \to \mathbb{N}$ is a computable permutation, then the set of pairs $\langle x, y \rangle$
for which $\langle x, i(y) \rangle \in U$ is a Gödel universal set as well. Therefore,
the theorem about isomorphism of Gödel numberings of enumerable
sets (p. 35) implies the following corollary: for any computable
permutation i, there exists a computable permutation i' such that

$$\langle x, y \rangle \in U_1 \Leftrightarrow \langle i'(x), i(y) \rangle \in U_2.$$

If i' luckily coincides with i, then i is the desired permutation. But
we want to replace luck by a reference to the Fixed Point Theorem.
There are a number of obstacles on this path, but all of them can be
overcome and now we are going to briefly explain how.

First of all, we must recall the proof of the theorem about the isomorphism of Gödel numberings to see that the function i' (or rather,
its number) is algorithmically constructible from the function i.
Then, we want to refer to the Fixed Point Theorem, but the problem
is that the corresponding construction assumes that i is a bijection.
Therefore, we must modify the construction in the isomorphism theorem so as to make it applicable to any computable function i. Not
only that, it must always yield some bijection i' for any computable
function i; then a fixed point will automatically be a bijection.

Let us formulate the corresponding generalization of the theorem
about the isomorphism of Gödel numberings. It will be preceded by
an auxiliary definition. Let $I\colon \mathbb{N} \to \mathbb{N}$ be an arbitrary function. We
say that a set A I-corresponds to a set B if either $B = I(A)$ (B is the
image of A under the mapping I) or $A = I^{-1}(B)$ (A is the preimage
of B under this mapping). (If I is a bijection, the two properties are
equivalent.)

Let U_1 and U_2 be two arbitrary Gödel numberings of enumerable
sets, and let $I\colon \mathbb{N} \to \mathbb{N}$ be a computable function. Then there
exists a computable bijection $i'\colon \mathbb{N} \to \mathbb{N}$ such that for any k, the set
with number k in the numbering U_1 I-corresponds to the set with
number $i'(k)$ in the numbering U_2.

This generalization of the isomorphism theorem is proved in exactly the same way as the theorem itself, and the number of the function i' can be obtained from the number of the function I effectively.

This enables us to apply the plan described above and find the function I such that $i' = I$ (so I is a bijection), which was our goal. $\square$

Problem 41. Complete this argument.

A similar theorem holds for Gödel universal functions.

Theorem 30. *Let $F_1, F_2 \colon \mathbb{N} \to \mathbb{N}$ be two Gödel universal functions for the class of all unary computable functions. Then there exists a computable permutation i such that*

$$F_1(x, y) = z \Leftrightarrow F_2(i(x), i(y)) = i(z)$$

for any natural x, y, and z.

Problem 42. Conduct the proof of this theorem similarly to the previous one, using Rogers' Theorem about the isomorphism of Gödel numberings (p. 34, Theorem 23).

Chapter 6

m-Reducibility and Properties of Enumerable Sets

1. m-reducibility

We have already encountered the following technique: to prove the undecidability of a certain set X (for instance, the set of all numbers of all functions with nonempty domain), we showed that if X were decidable, then any enumerable set K would be decidable as well. This was done by the "reduction" argument: we constructed a total computable function f such that the membership of an arbitrary number n in the set K was determined by the membership of the number $f(n)$ in the set X.

Now we will study these situations in more detail.

We say that a set A of natural numbers is *m-reducible* to another set B of natural numbers if there exists a total computable function $f \colon \mathbb{N} \to \mathbb{N}$ such that

$$x \in A \Leftrightarrow f(x) \in B$$

for all $x \in \mathbb{N}$. We say that the function f *m-reduces* A to B. Notation: $A \leq_m B$.

Theorem 31. (a) *If* $A \leq_m B$ *and* B *is decidable, then* A *is decidable.* (b) *If* $A \leq_m B$ *and* B *is enumerable, then* A *is enumerable.* (c) $A \leq_m A$; *if* $A \leq_m B$ *and* $B \leq_m C$, *then* $A \leq_m C$. (d) *If* $A \leq_m B$, *then* $\mathbb{N} \setminus A \leq_m \mathbb{N} \setminus B$.

Proof. All these properties are almost obvious. Suppose that $A \leq_m B$ and we have a deciding algorithm for B. To find out whether a given x belongs to the set A, we compute $f(x)$ and check if $f(x)$ belongs to B. In other words, we can write $a(x) = b(f(x))$, where a is the characteristic function of the set A and b is the characteristic function of B; therefore, if b is computable, then a is also computable as the composition of computable functions.

A similar equation can be written for semicharacteristic functions; therefore, the enumerability of B implies the enumerability of A. Another way to say this: the set A is the preimage of the enumerable set B under the computable mapping f, hence A is enumerable.

Obviously, the identity function m-reduces A to A. If a function f reduces A to B and g reduces B to C, then

$$x \in A \Leftrightarrow f(x) \in B \Leftrightarrow g(f(x)) \in C,$$

so the composition of the functions g and f reduces A to C.

Finally, any function reducing A to B will also reduce $\mathbb{N} \setminus A$ to $\mathbb{N} \setminus B$. □

Historically, the letter "m" comes from the expression "many-one-reducibility"; however, as Michael Sipser suggests in his "Introduction to the Theory of Computation" [**14**], we can say "mapping reducibility" retaining the letter m in the notation.

It should be mentioned that this definition is not symmetric with respect to taking the complement of only one of the sets in the relation: it is not at all necessary that $A \leq_m \mathbb{N} \setminus A$, although we always have $A \leq_m A$.

Problem 43. Show that $A \not\leq_m \mathbb{N} \setminus A$ for an enumerable undecidable set A.

Notice that the sets $\varnothing$ and $\mathbb{N}$ are special cases with respect to m-reducibility. For example, any decidable set A is reducible to any set B unless B is empty or coincides with $\mathbb{N}$. Indeed, if $p \in B$, $q \notin B$, and A is decidable, then the reducing function can be constructed as follows:

$$f(x) = \textbf{if } x \in A \textbf{ then } p \textbf{ else } q \textbf{ fi}.$$

But if B is empty or coincides with $\mathbb{N}$, then only the empty set (or $\mathbb{N}$, respectively) is m-reducible to B.

Problem 44. Does there exist a set of natural numbers such that any set of natural numbers is m-reducible to it?

2. *m*-complete sets

Theorem 32. *In the class of enumerable sets, there are sets maximal with respect to m-reducibility, that is, sets to which any enumerable set can be m-reduced.*

Proof. It suffices to consider any universal set (formally, we must replace pairs by their numbers). Indeed, let $U \subset \mathbb{N} \times \mathbb{N}$ be an enumerable set of pairs of natural numbers which is universal for the class of enumerable sets of natural numbers. Consider the set V of numbers of all pairs from U (under any computable numbering $\langle x, y \rangle \leftrightarrow [x, y] \in \mathbb{N}$ of the pairs):

$$V = \{[x, y] \mid \langle x, y \rangle \in U\}.$$

Let T be an arbitrary enumerable set. Then $T = U_n$ for a certain n, and hence

$$x \in T \Leftrightarrow x \in U_n \Leftrightarrow \langle n, x \rangle \in U \Leftrightarrow [n, x] \in V.$$

Thus the function $x \mapsto [n, x]$ reduces T to V. □

Enumerable sets maximal with respect to m-reducibility are called *m-complete* (more exactly, *m-complete in the class of enumerable sets*).

Notice that if $K \leq_m A$, where K and A are enumerable sets, and K is m-complete, then A is also m-complete (by transitivity).

The diagonal of a Gödel universal set is m-complete:

Theorem 33. *Let $U \subset \mathbb{N} \times \mathbb{N}$ be a Gödel universal set for the class of enumerable sets. Then its "diagonal section" $D = \{x \mid \langle x, x \rangle \in U\}$ is m-complete.*

(In particular, the set of all self-applicable programs is m-complete.)

Proof. Obviously, D is enumerable. Let K be an arbitrary enumerable set. Consider the enumerable set of pairs $V = K \times \mathbb{N}$. Its sections V_n are either empty (for $n \notin K$) or coincide with the entire set $\mathbb{N}$ (for $n \in K$).

Since U is a Gödel set, there exists a total computable function s such that $V_n = U_{s(n)}$. In other words, $U_{s(n)}$ coincides with $\mathbb{N}$ for $n \in K$ and is empty for $n \notin K$. Consequently, $s(n) \in U_{s(n)}$ (and so $s(n) \in D$) for $n \in K$, and $s(n) \notin U_{s(n)}$ (and so $s(n) \notin D$) for $n \notin K$. It follows that s reduces K to D. □

Problem 45. Prove that the set of all programs that halt on the input 0 is m-complete. Prove that the set of all programs that halt on at least one input is m-complete.

Problem 46. Let M be an m-complete enumerable set. Show that there exists an algorithm that transforms any number of any total function h into an integer n such that $(n \in M) \Leftrightarrow (h(n) \in M)$. (*Hint*: in essence, this statement is the Fixed Point Theorem for some equivalence relation.)

3. m-completeness and effective nonenumerability

The theory of algorithms makes it possible, so to speak, to "constructivize" various definitions. For example, consider the definition of an infinite set. What is an infinite set? It is a set that contains at least n elements for any natural n. Now we can say that a set is called "effectively infinite" if there exists an algorithm that returns n distinct elements of this set for any given n.

Problem 47. Show that an arbitrary set A is effectively infinite if and only if it contains an infinite enumerable set (i.e., it is not immune; see p. 17).

Now let us consider the effective version of the notion of nonenumerability. What does it mean to say that a set A is not enumerable? This simply means that A differs from any enumerable set. So it is natural to call a set A effectively nonenumerable if for any enumerable set we can show a "place" where it differs from A, that is, a natural number that belongs exactly to one of the two sets.

More formally, let us fix a Gödel universal enumerable set W (and thus a numbering of enumerable sets: any natural n is considered to be a number of the set W_n). We will say that a set A is *effectively nonenumerable* if there exists a total computable function d such that $d(z) \in A \triangle W_z$ for all z. (Here $\triangle$ denotes symmetric difference; in other words, $d(z)$ is a point where A differs from W_z.)

Notice that this property does not depend on the choice of the Gödel universal set, because we can effectively pass from the numbers with respect to one of these sets to the numbers with respect to any other.

The property of effective nonenumerability can be characterized in terms of m-reducibility. We start with the following simple observation.

Theorem 34. *If $A \leq_m B$ and A is effectively nonenumerable, then B is also effectively nonenumerable.*

Proof. This theorem is the "effective version" of Theorem 31, part (b). The same is true for its proof. Suppose that we want to find a point at which B differs from a given enumerable set X. Consider the function f that m-reduces A to B. The preimage $f^{-1}(X)$ of the enumerable set X under a computable map f is enumerable; therefore, we can find a point m at which it differs from A. Then B differs from X at the point $f(m)$.

To complete this argument, we must prove that one can effectively obtain a number of the enumerable set $f^{-1}(X)$ from a number of the enumerable set X. This proof must involve the fact that we use a *Gödel* numbering: we will follow the same lines as in our computation of a number of the composition of two computable functions specified by their numbers (Theorem 16). Here are the details.

Consider the enumerable set

$$V = \{\langle x, y \rangle \mid \langle x, f(y) \rangle \in W\}$$

where W is a Gödel universal set that defines our numbering. The set V is enumerable as the preimage of the enumerable set W under the computable map $\langle x, y \rangle \mapsto \langle x, f(y) \rangle$. It is easy to see that $V_n = f^{-1}(W_n)$. Since W is a Gödel universal set, there exists a total computable function s such that $W_{s(n)} = V_n = f^{-1}(W_n)$ for all n. In other words, the function s maps a W-number of any enumerable set into a W-number of its preimage under the map f. The proof is complete. □

Theorem 35. *There exist enumerable sets with effectively nonenumerable complements.*

Proof. Consider the diagonal set $D = \{n \mid \langle n, n \rangle \in W\}$ again. Its complement is effectively nonenumerable. Indeed, the sets W_n and D do not differ at the point n (they both contain or do not contain n simultaneously), so W_n differs from the complement of D at n. It follows that the complement of D is effectively nonenumerable: we can set the function d in the definition of effective nonenumerability to be simply the identity function. □

The two previous theorems obviously imply the following statement.

Theorem 36. *The complement of any m-complete enumerable set is effectively nonenumerable.*

In fact, the converse statement is also true, as the following theorem shows:

Theorem 37. *Let K be an enumerable set, and let A be effectively nonenumerable. Then $\mathbb{N} \setminus K \leq_m A$ (or, equivalently, $K \leq_m \mathbb{N} \setminus A$).*

Proof. What really counts here is only the ability to effectively distinguish A from two enumerable sets: the empty set and the entire set $\mathbb{N}$. To distinguish A from the empty set means to specify an element in A; to distinguish A from the entire set of natural numbers means to specify an element outside A. It is these two observations on which the reduction is based. More formally, consider the set $V = K \times \mathbb{N}$. Its sections V_n are either empty (for $n \notin K$) or coincide with $\mathbb{N}$ (for $n \in K$). Using the fact that W is a Gödel set, we find a total function s such that $W_{s(n)} = \varnothing$ for $n \notin K$ and $W_{s(n)} = \mathbb{N}$ for $n \in K$. Let d be the function that ensures the effective nonenumerability of the set A. Then $d(s(n)) \in A$ for $n \notin K$ and $d(s(n)) \notin A$ for $n \in K$. In other words, the composition of the functions d and s reduces $\mathbb{N} \setminus K$ to the set A, completing the proof. $\qquad\square$

Hence we have the following obvious corollaries.

Theorem 38. *An enumerable set is m-complete if and only if its complement is effectively nonenumerable.*

Theorem 39. *A set is effectively nonenumerable if and only if the complement of some (another version: any) m-complete set is m-reducible to it.*

Notice that not all nonenumerable sets are effectively nonenumerable. This can be deduced, e.g., from the following fact.

Theorem 40. *Any effectively nonenumerable set contains an infinite enumerable subset (i.e., is not immune).*

Proof. Indeed, suppose that A is an effectively nonenumerable set. We can find a point where A differs from the empty set, that is, an element in A. Then we spot its distinction from the one-element set consisting of this element; thus we get another element of A. Proceeding the same way, we can algorithmically find arbitrarily many pairwise distinct elements.

This reasoning implicitly uses the following fact: given a finite set specified by the list of its elements, we can obtain (some) number of this set in a Gödel numbering of enumerable sets. Why is this true? Let us fix a certain computable numbering of finite sets. Denote by D_n the nth finite set in this numbering. Then D_n is the nth section of the enumerable (and even decidable) set

$$D = \{\langle n, x \rangle \mid x \in D_n\}.$$

It remains to apply the definition of a Gödel numbering of enumerable sets. $\qquad\square$

This theorem shows that simple sets (which exist according to Theorem 14) are enumerable sets that are not m-complete. In fact, the notion of simple sets was introduced exactly for this reason: Post was looking for an example of an enumerable and undecidable, but not m-complete set.

4. Isomorphism of m-complete sets

In this section we will prove that all m-complete sets "have the same structure" and differ from one another only by a computable permutation.

Theorem 41. *Let A and B be m-complete enumerable sets. Then there exists a computable permutation (a computable one-to-one correspondence) $f \colon \mathbb{N} \to \mathbb{N}$ that maps A onto B, that is, $x \in A \Leftrightarrow f(x) \in B$ for all x.*

Proof. We will use the same technique as in the proof of Rogers' theorem about isomorphism of Gödel numberings (see p. 34). To begin with, we will prove the following lemma.

Lemma. *Let A be an m-complete enumerable set. Then it is possible to algorithmically obtain from any natural number n arbitrarily many other natural numbers that are A-equivalent to n (we say that k and l are A-equivalent if either $k \in A$ and $l \in A$ or $k \notin A$ and $l \notin A$).*

Proof of the Lemma. As we did before, we will use two methods to obtain new numbers that are A-equivalent to a given number n. One method will succeed (give a new number) if $n \in A$, the other if $n \notin A$. We will apply both methods unaware of which of the two possibilities is actually the case (moreover, we may never find out which).

The first method: Let P be an enumerable undecidable set. Consider an enumerable set of pairs $A \times P$. It is m-reducible to A, since A is m-complete. (The definition of reducibility deals with sets of natural numbers rather than sets of pairs, but as usual this does not matter, because we can replace pairs by their numbers.) In other words, there exists a total computable function f of two natural arguments with the following property:

$$f(n, m) \in A \Leftrightarrow (n \in A) \text{ and } (m \in P).$$

In particular, for $m \in P$, numbers n and $f(n,m)$ are A-equivalent. Therefore, having arranged P into a computable sequence $p(0), p(1), \ldots$, we can compute the numbers $f(n, p(0)), f(n, p(1)), \ldots$ and obtain new numbers that are A-equivalent to n.

Suppose that $n \in A$. Let us show that the set X of the numbers obtained by this method is infinite (in this case all these numbers belong to A as well). If $m \in P$, then $f(n,m) \in X$ (by the construction of X), and if $m \notin P$, then $f(n,m) \notin X$ (since in this case $f(n,m) \notin A$, and $X \subset A$). Thus the function $m \mapsto f(n,m)$ reduces the undecidable set P to the set X; therefore, X is undecidable, and hence infinite.

Now let us describe the second method, which works in the case $n \notin A$. Let us take two enumerable inseparable sets P and Q. Consider the enumerable set of pairs $(A \times P) \cup (\mathbb{N} \times Q)$. Denote by f the function reducing it to A. This means that $f(n,m) \in A$ if and only if $(n \in A$ and $m \in P)$ or $m \in Q$. As before, for $m \in P$, the numbers n and $f(n,m)$ are A-equivalent, so we can consider the sequence $f(n, p(0)), f(n, p(1)), \ldots$ again; it only remains to show that (for $n \notin A$) this sequence contains infinitely many distinct terms.

Suppose that this is not true and the set X of all terms of this sequence is finite. By our assumption, X is disjoint with A. Notice that $f(n,m) \in X$ if $m \in P$ (by construction) and $f(n,m) \notin X$ if $m \in Q$ (since in this case $\langle n, m \rangle$ belongs to our enumerable set of pairs and $f(n,m)$ belongs to A). Thus the preimage of the set X under the mapping $m \mapsto f(n,m)$ separates P from Q. But this preimage is decidable (X is decidable as a finite set, and the mapping in question is defined everywhere and computable), whereas by our assumption the sets P and Q cannot be separated by a decidable set.

So we have described two methods of generating new numbers that are A-equivalent to the given one. Their parallel application will inevitably give the desired result. This completes the proof of the Lemma.

Now suppose that A and B are two m-complete enumerable sets. Let us prove that they differ only by a computable permutation of natural numbers. We will construct this permutation step by step.

At the kth step we have a one-to-one correspondence

$$a_1 \leftrightarrow b_1, a_2 \leftrightarrow b_2, \ldots, a_k \leftrightarrow b_k$$

such that $a_i \in A \Leftrightarrow b_i \in B$ for all i. At even steps, we take the smallest number not included in the left-hand side of this correspondence. Using the m-reducibility of A to B we find a counterpart to this number. In so doing, the Lemma allows us to choose the counterpart that has not yet appeared among the elements on the right-hand side. At odd steps, we do the same, but from right to left.

In the limit, this process yields the desired computable permutation that links A and B. □

From the standpoint of the theory of algorithms, two sets that differ only by a computable permutation have the same properties. Therefore, the theorem proved above shows that, in essence, there is only one m-complete enumerable set (or, equivalently, only one enumerable set with effectively nonenumerable complement).

5. Productive sets

In this section we use the Fixed Point Theorem to obtain the following unexpected, at first sight, result: the definition of effective nonenumerability of a set A will not change if we confine ourselves only to (enumerable) subsets of the set A.

Let us fix some Gödel numbering of enumerable sets (the set with number n is denoted by W_n). A set A is called *productive* if there exists a computable (but not necessarily total) function f with the following property: for any n such that $W_n \subset A$ the value $f(n)$ is defined and $f(n) \in A \setminus W_n$.

Problem 48. Prove that a productive set cannot be immune.

Clearly, the requirements in the definition of productivity constitute only a part of those in the definition of an effectively nonenumerable set, so any effectively nonenumerable set is productive. However, as surprising as it may seem, the converse statement turns out to be true as well.

Theorem 42. *Let A be a productive set, and let K be an arbitrary enumerable set. Then the complement of K is m-reducible to A.*

(As we have seen above, in this case A is effectively nonenumerable.)

Proof. Let f be the function from the definition of productivity (the one that yields an element outside the subset with a given number).

We will construct a total computable function s with the following properties:

- $x \notin K \Rightarrow W_{s(x)} = \varnothing$;
- $x \in K \Rightarrow W_{s(x)} = \{f(s(x))\}$.

(The second property implies that $f(s(x))$ is defined for $x \in K$.) Before we proceed to the construction, which uses the Fixed Point Theorem, let us notice that in the first case $f(s(x))$ is defined and belongs to A: since the set with the number $s(x)$ is empty, and the empty set is a subset of A, the number $f(s(x))$ must be an element of A. On the contrary, in the second case $f(s(x))$ does not belong to A. Indeed, suppose that $f(s(x)) \in A$. Then the set $W_{s(x)}$ is a subset of A; hence the number $f(s(x))$ is an element of A that does not belong to this subset, whereas it does.

Therefore, if we manage to construct such a function s, then the function $x \mapsto f(s(x))$ will be a total computable function that m-reduces the complement of the set K to the set A, as stated in the theorem. So how do we construct s?

If in the second property (for the case $x \in K$) we had, say, $f(x)$ instead of $f(s(x))$, then we would have no problem. As usual, in this case we can consider the enumerable set of pairs

$$V = \{\langle x, y \rangle \mid x \in K \text{ and } y = f(x)\};$$

sections of this set are of the desired form, and we only have to use the fact that the numbering is Gödel. Coming back to our actual goal, with $f(s(x))$ in the right-hand side of the second property, we see that, as in the classical question about the hen and the egg, $s(x)$ is needed to construct V, and V is needed to construct $s(x)$.

But this is exactly the sort of difficulty that the Fixed Point Theorem helps to overcome. Let us construct a binary total computable function h with the following properties:

- $x \notin K \Rightarrow W_{h(x,t)} = \varnothing$;
- $x \in K \Rightarrow W_{h(x,t)} = \{f(t)\}$.

(We have done similar things many times, the last time in the previous paragraph. Notice that $f(t)$ can be undefined; then by $\{f(t)\}$ we mean the empty set.) By the Fixed Point Theorem (for enumerable sets), for each x the function $t \mapsto h(x,t)$ has a fixed point, and as we have mentioned in the section about the fixed point with parameter, this fixed point can be chosen computably depending on x. Thus there exists a total computable function s such that

$$W_{s(x)} = W_{h(x,s(x))}$$

for all x. This equation can be extended:

$$W_{s(x)} = W_{h(x,s(x))} = \begin{cases} \varnothing \text{ if } x \notin K, \\ \{f(s(x))\} \text{ if } x \in K; \end{cases}$$

and this is exactly what we wanted. Notice that the value $f(s(x))$ is defined for all x (if it were undefined, then $W_{s(x)} = \varnothing$, but in this case $f(s(x))$ must be defined by the definition of productivity). Thus the Fixed Point Theorem yields a concordant "egg-hen" couple, completing the proof.   □

Enumerable sets whose complements are productive are called *creative*. The name derives from the fact that such a set (to be more exact, its complement) is more "inventive" than any algorithmic process: whatever a method is for generating elements in the complement, one can find an element in the complement that cannot be obtained by this method.

As we see, creative sets, enumerable sets with effectively nonenumerable complements, and m-complete sets form the same class, and any two sets in this class differ only by a computable permutation.

Elements of a productive set can be generated by the following inductive process. We start with the empty set. Applying to it the productive function (i.e., the function mentioned in the definition of a productive set), we obtain some element. It constitutes a one-element subset. Applying the productive function to this subset, we obtain another element. Then we can apply the productive function to the two-element subset thus obtained again, and so on. The process yields an infinite computable sequence of elements of the given productive set. (We have already done this when we proved that an effectively nonenumerable set contains an infinite enumerable subset.) But this is not the end: the inductive process can be "transfinitely" continued: given an enumerable subset of our productive set (the set of all terms of the sequence), we can find another element of the productive set (the element number ω, so to speak). We add it to the sequence, apply the productive function again to obtain the $(\omega + 1)$th element, and so on; then a new sequence arises, then an $(\omega \cdot 2)$nd element, an $(\omega \cdot 3)$rd,..., an ω^2th element, etc.

But of course, it is impossible to obtain an algorithm enumerating a productive (and hence nonenumerable) set.

Problem 49. Without using the Fixed Point Theorem (and Theorem 42), show that for any productive set A there exists a *total* computable function f such that $W_n \subset A$ implies $f(n) \in A \setminus W_n$. (*Hint*: alternate W_n with the empty set, as we did in the proof of the Lemma to Theorem 41.)

6. Pairs of inseparable sets

In this section we formulate a number of results concerning pairs of disjoint enumerable sets. These results are parallel to the theorems about m-completeness, productivity, effective nonenumerability, and isomorphism of m-complete sets proved above.

Let A and B be two disjoint sets (of natural numbers). Recall that they are called inseparable if there is no decidable set containing one of them and disjoint with the other. This definition can be reformulated as follows: if W_x and W_y are two disjoint enumerable sets containing A and B, respectively, then the union $W_x \cup W_y$ does not contain all natural numbers. (It will be convenient to denote enumerable sets by W_x and W_y, assuming that W is a Gödel universal set.)

Now it becomes clear how an effective version of this definition can be formulated. We say that disjoint sets A and B are *effectively inseparable* if there exists a computable function h such that if $A \subset W_x$, $B \subset W_y$, and $W_x \cap W_y = \varnothing$, then $h(x, y)$ is defined and $h(x, y) \notin W_x \cup W_y$.

The definition of inseparability can be formulated in a slightly different way: there is no total computable function φ_n equal to zero at any point of the set A and equal to one at any point of the set B. (We assume that φ is a Gödel universal function.) The effective version changes accordingly: sets A and B are *strongly effectively inseparable* if there exists a total computable function h that for any n returns a point $h(n)$ at which the function φ_n "errs". There are three possible error types: (1) $\varphi_n(h(n))$ is undefined, (2) $h(n) \in A$, but $\varphi_n(h(n))$ is not equal to zero, or (3) $h(n) \in B$, but $\varphi_n(h(n))$ is not equal to one.

Problem 50. Show that strong effective inseparability implies effective inseparability (justifying our terminology).

The converse statement is also true, but its proof is more complicated, and we will return to it later.

Do strongly effectively inseparable enumerable sets exist? It is readily seen that the standard diagonal construction yields a pair of such sets, namely, the sets $\{x \mid \varphi_x(x) = 1\}$ and $\{x \mid \varphi_x(x) = 0\}$, for which we can take the identity function as the function h.

Problem 51. Check this.

Extending our analogy between sets and pairs, let us define the notion of m-reducibility for pairs. Again, we will have two versions. Let $\langle A, B \rangle$ and $\langle C, D \rangle$ be two pairs of disjoint enumerable sets ($A \cap B = C \cap D = \varnothing$). We say that a total computable function f m-*reduces* $\langle A, B \rangle$ *to* $\langle C, D \rangle$ if $f(A) \subset C$ and $f(B) \subset D$.

Problem 52. (a) Show that if f reduces $\langle A, B \rangle$ to $\langle C, D \rangle$ and C can be separated from D by a decidable set, then A can be separated from B by a decidable set as well. (b) Show that if f reduces $\langle A, B \rangle$ to $\langle C, D \rangle$ and the pair $\langle A, B \rangle$ is effectively inseparable, then the pair $\langle C, D \rangle$ is effectively inseparable as well. (c) Show that if f reduces $\langle A, B \rangle$ to $\langle C, D \rangle$ and the pair $\langle A, B \rangle$ is strongly effectively inseparable, then the pair $\langle C, D \rangle$ is strongly effectively inseparable as well.

The definition of reducibility can be strengthened by the additional requirement that $f(x) \notin C \cup D$ for $x \notin A \cup B$ (in other words, f must simultaneously reduce A to C and B to D). In this case we will say that f *strongly reduces* the pair $\langle A, B \rangle$ to the pair $\langle C, D \rangle$.

Now we can define *m-completeness* and *strong m-completeness* for a pair of disjoint enumerable sets by requiring that any pair of disjoint enumerable sets is m-reducible (strongly m-reducible) to the given one.

Problem 53. Show that if a pair is strongly effectively inseparable, then it is strongly m-complete. (*Hint.* Let a pair $\langle A, B \rangle$ be strongly effectively inseparable, and let $\langle K, L \rangle$ be any pair of disjoint enumerable sets. For any natural number x we can construct a computable function ψ_x with the following properties: if $x \in K$, then ψ_x is total and differs from 1 only at finitely many points, all of them in A; if $x \in L$, then ψ_x is total and differs from 0 only at finitely many points, all of them in B; and if $x \notin K \cup L$, then ψ_x is equal to 0 on A and to 1 on B. To construct this function, we enumerate K and L, adding to the graph of ψ_x pairs of the form $\langle a, 0 \rangle$ and $\langle b, 1 \rangle$ until the element x crops up in one of these sets; as soon as this happens, we declare ψ_x to be a corresponding constant (0 or 1) at all other points. Then it remains to use properties of the Gödel numbering φ and strong effective inseparability of A and B.)

Problem 54. Show that any m-complete pair is strongly effectively inseparable. (*Hint:* a strongly effectively inseparable pair exists and is reducible to the pair in the exercise.)

From the statements of these exercises, it follows that the properties of m-completeness, strong m-completeness, and strong effective inseparability are equivalent. It can be proved that the property of effective inseparability, which seems to be weaker, is in fact also equivalent to them. The proof is similar to that of Theorem 42 (which establishes the m-completeness of any creative set). Notice that the difference between effective inseparability and strong effective inseparability is more or less the same as between productivity and effective nonenumerability.

Problem 55. Let $\langle A, B \rangle$ be an effectively inseparable pair of disjoint sets. Show that it is strongly m-complete. (*Hint.* Let K and L be arbitrary disjoint enumerable sets. Let h be the function from the definition of effective inseparability (of the sets A and B). Using the Fixed Point Theorem, construct total computable functions $x(n)$ and $y(n)$ with the following properties: (1) if $n \in K$, then $W_{x(n)} = A$, $W_{y(n)} = B \cup \{h(x(n), y(n))\}$; (2) if $n \in L$, then $W_{x(n)} = A \cup \{h(x(n), y(n))\}$, $W_{y(n)} = B$; (3) if $n \notin K \cup L$, then $W_{x(n)} = A$, $W_{y(n)} = B$. Then show that for $n \in K$, the value $h(x(n), y(n))$ is defined and belongs to A; for $n \in L$, the value $h(x(n), y(n))$ is defined and belongs to B; and for $n \notin K \cup L$ the value $h(x(n), y(n))$ is defined and lies outside $A \cup B$.)

Thus all the four properties formulated above are equivalent. Extending our analogy, we can show that any two pairs of effectively inseparable sets are isomorphic; to this end, as a prerequisite, we must learn how to obtain arbitrarily many numbers "equivalent" to a given number with respect to a pair of effectively inseparable sets.

More exactly, suppose that A and B are disjoint sets. We say that two numbers are $\langle A, B \rangle$-*equivalent* in any of the following three cases: they both belong to A, they both belong to B, or they both do not belong to $A \cup B$. (So we have three equivalence classes: the set A, the set B, and the remainder.)

Problem 56. Let $\langle A, B \rangle$ be a strongly m-complete pair of enumerable sets. Show that from any number k we can algorithmically obtain arbitrarily many pairwise distinct numbers that are $\langle A, B \rangle$-equivalent to k. (*Hint*: proceed by analogy with the proofs of Theorem 22 and the Lemma to Theorem 41.)

Problem 57. Let $\langle A_1, B_1 \rangle$ and $\langle A_2, B_2 \rangle$ be two strongly m-complete pairs of enumerable sets. Then they are computably isomorphic in the following sense: there exists a computable permutation (bijection) $i \colon \mathbb{N} \to \mathbb{N}$ such that $i(A_1) = A_2$ and $i(B_1) = B_2$. (*Hint*: proceed by analogy with the proofs of Theorems 23 and 41.)

Chapter 7

Oracle Computations

1. Oracle machines

If a set B is m-reducible to a decidable set A, then B is also decidable. Furthermore, even if A is undecidable, but we have access to an "oracle" for A that answers questions about the membership of numbers in the set A, then we can use it to answer questions about the membership of numbers in the set B. Indeed, if f is the reducing function and if we want to find out whether a certain number x belongs to B, it suffices to ask the oracle whether $f(x)$ belongs to A.

Clearly, m-reducibility employs the potential of the oracle in a rather limited way: first, only one question is asked; second, the answer to this question is taken as the answer to the original question about the membership of a number x in the set B. Here is an example that does not fit in this scheme: given an oracle for the set A, we can answer questions about the membership of numbers in the set $B = \mathbb{N} \setminus A$. As before, only one question is asked and answered, but the answer to this question is to be reversed. Another example: given an oracle for the set A, we can answer questions about the membership of a pair of natural numbers in the set $B = A \times A$. (Here the oracle must be asked two questions.)

Therefore, it is natural to look for a more general definition of reducibility of a set B to a set A. Let us say that B is reducible to A if there exists an algorithm that decides the set B using an "oracle" answering questions about the set A. In other terms: there exists an algorithm that makes calls to an external function a(x:integer):boolean; this algorithm decides the set B if the calls to a(x) return correct values of the expression "$x \in A$" (i.e., return true if $x \in A$ and false if $x \notin A$). If B is reducible to A, we also say that B is A-*decidable*.

This kind of reducibility is called *Turing reducibility*, or T-reducibility. Notation: $B \leq_T A$ means that B is Turing-reducible to A. Here are a few simple facts about T-reducibility:

Theorem 43. (a) *If* $B \leq_m A$, *then* $B \leq_T A$. (b) $A \leq_T \mathbb{N} \setminus A$ *for any* A. (c) *If* $A \leq_T B$ *and* $B \leq_T C$, *then* $A \leq_T C$. (d) *If* $A \leq_T B$ *and* B *is decidable, then* A *is decidable.*

Proof. All these statements are almost obvious. Let us explain, e.g., statement (c). Suppose that we have an algorithm for A that includes calls to an external decision procedure for B, as well as an algorithm for B that includes calls to an external procedure for C. Then we can replace the calls to the external B-procedure by the second algorithm to obtain a deciding algorithm for A that uses calls to the external procedure for C.   □

Notice that a nonenumerable set can be T-reducible to an enumerable one. (This is impossible for m-reducibility.) For instance, the complement of an enumerable undecidable set K is reducible to K.

We have defined the notion of an A-decidable set. We can define the notion of an A-computable function in a similar way: a function f is A-computable if there exists an algorithm M (with calls to an oracle) that computes f if these calls are correctly answered by the A-oracle. Recall that this means that if $f(x)$ is defined, then the algorithm terminates and returns $f(x)$ on the input x, and if $f(x)$ is undefined, then it does not halt. In this case the (partial) function f is said to be *computable relative to the set A* or A-*computable*.

In our definition of reducibility, the external function takes only two values ("yes" and "no"). This restriction is not essential. Let $\alpha \colon \mathbb{N} \to \mathbb{N}$ be an arbitrary total function. Then we can speak about functions *computable relative to* α; the algorithms that compute them include calls of a function α. However, this generalization does not give us anything new, as the next theorem shows.

Theorem 44. *A partial function f is computable relative to a total function α if and only if it is computable relative to the graph of the function α, that is, relative to the set $\{\langle n, \alpha(n) \rangle \mid n \in \mathbb{N}\}$.*

Proof. Indeed, if we can call a function α, then we can answer questions about the membership of an arbitrary pair in the graph of α. Conversely, by using a decision procedure for the graph of α as an oracle, we can find $\alpha(x)$ for a given x successively asking questions about the membership of pairs $\langle x, 0 \rangle, \langle x, 1 \rangle, \ldots$ in the graph until we get a positive answer. $\square$

In the definition of α-relative computability, we assumed that the function α is total. This restriction is fundamental: the semantics of calling nontotal functions is not well defined. Suppose that we have called $\alpha(x)$ and it turned out that the function α is not defined at x. Does this mean that the algorithm "hangs up" (with no output)? Or that we can launch a parallel computation and produce a result before the answer $\alpha(x)$ arrives? Can we request several values of the function α in parallel? For instance, is the function $f(x)$ defined by the formula

$$f(x) = \begin{cases} 0 \text{ if } \alpha(2x) \text{ or } \alpha(2x+1) \text{ is defined,} \\ \text{undefined otherwise} \end{cases}$$

computable relative to α? As we see, there are different (in fact, nonequivalent) versions of definitions and each of them has its own deficiencies. To avoid these problems, we will consider α-computability only for total functions α.

Problem 58. Suppose that we have two different sets X and Y. Let us consider programs that have access to two oracles, for X and for Y, and functions that can be computed by means of these programs.

Show that this definition adds nothing essentially new: there exists a set Z such that X-Y-computability coincides with Z-computability.

2. Relative computability: Equivalent description

Now we are going to give an equivalent definition of α-computable functions that does not involve oracle calls.

Recall that a function with natural values defined on a finite subset of the set of natural numbers is called a pattern. A pattern is defined by a list of pairs $\langle \text{argument}, \text{value} \rangle$. Patterns can be computably numbered; after that, we can identify a pattern and its number and speak about a decidable set of patterns, enumerable set of patterns, etc.

Two patterns are called *coherent* if the union of their graphs is the graph of a function, that is, if there is no point at which both of them are defined and take different values.

Suppose that M is a set of triples of the form $\langle x, y, t \rangle$, where x and y are natural numbers and t is a pattern. We will say that two triples $\langle x_1, y_1, t_1 \rangle$ and $\langle x_2, y_2, t_2 \rangle$ are *incompatible* if the patterns t_1 and t_2 are coherent, $x_1 = x_2$, but $y_1 \neq y_2$. A set M will be called *consistent* if it does not contain incompatible triples.

Let M be a consistent set, and let α be some function. Consider all the triples $\langle x, y, t \rangle \in M$ such that t is a part of α (i.e., the graph of t is a subset of the graph of α). All patterns in the chosen triples are coherent; therefore, since M is consistent, no two of the chosen triples can have equal first and different second elements. This means that by omitting the third elements of the chosen triples, we obtain the graph of some function (generally speaking, a partial one). This function will be denoted by $M[\alpha]$.

Theorem 45. *A partial function $f \colon \mathbb{N} \to \mathbb{N}$ is computable relative to a total function $\alpha \colon \mathbb{N} \to \mathbb{N}$ if and only if there exists an enumerable consistent set of triples M such that $f = M[\alpha]$.*

Proof. Suppose that the function f is computed by a program p with calls to the external procedure α. For each natural number x, we simulate the program operation on the input x along all paths, that is, taking into account that each call to α can return any value $\alpha(n)$. Then for each x a tree of possible computation paths emerges: each call to the external procedure is represented by a fork with countable branching. On some branches of the tree, the computation terminates and the program returns an answer. Once we find a branch that produces an answer y on the input x, we form a triple $\langle x, y, t \rangle$, where t is the pattern that contains all arguments and values of the function α used along this branch.

The set M of triples thus obtained is enumerable (the procedure described above allows us to generate all its elements one by one). This set does not contain incompatible triples. Indeed, if two triples $\langle x, y_1, t_1 \rangle$ and $\langle x, y_2, t_2 \rangle$ in this set have the same x and $y_1 \neq y_2$, then they correspond to different paths in the computation tree on the same input x. These paths diverge at a certain point; this means that at a certain step we have received different answers to the same question. These different answers were registered in the patterns t_1 and t_2, so the patterns are not coherent. Therefore, the set M is consistent.

Notice that the set M depends on the program p but not on the function α (we took into account all possible answers that could be given by the oracle, so we did not use α in the construction of M). Now we have to check that for any total function α, the function $M[\alpha]$ coincides with the function computed by the program p with α as an oracle.

Suppose that the program p computes the function f using the oracle α and $f(x) = y$, that is, the program p returns the answer y on the input x. The computation involves a number of calls to the function α and is represented by a certain branch of the tree considered above. Let t be the pattern that contains all the questions asked along this branch and the corresponding answers. Then t is a part of α. In addition, the triple $\langle x, y, t \rangle$ belongs to the set M. Hence $M[\alpha](x)$ is defined and is equal to y.

Conversely, if $M[\alpha](x) = y$, then there exists a triple $\langle x, y, t \rangle \in M$ such that t is a part of α. This triple corresponds to a certain branch of the computation tree. Since t is a part of α, the α-oracle computation follows this branch and the program returns the answer y.

Thus, for any program p, we have constructed an enumerable consistent set M that specifies the same function as the program p, which completes the proof of the "only-if" statement.

In order to prove the "if" statement, suppose that a consistent set M is given. We must construct a program p equivalent to M. This program, supplied with the oracle that computes α, will compute the function $M[\alpha]$. We define the program p as follows: having received the input x, it enumerates the set M and picks out the triples with the first element x. For each of these triples $\langle x, y, t \rangle$, it calls the external procedure (questions the oracle) to find out whether t is a part of the function α. If it is, the computation terminates and the program returns the answer y, if not, then the enumeration of M continues.

Obviously, the program p thus constructed computes the function $M[\alpha]$.   □

Problem 59. Suppose that we have carried out this construction in both directions: first, from a given consistent set M, we have constructed a program as described in the second part of the proof, and then, starting with this program, we constructed a consistent set M'. Can the two sets M' and M be distinct?

3. Relativization

Let us fix a total function α. Then the entire theory of computable functions can be, as they say, "relativized" with respect to α, by replacing computable functions in all definitions and statements by α-computable functions (i.e., functions computable relative to α). In so doing, all the results formulated in previous chapters remain valid, and their proofs undergo only slight changes.

In particular, we can define the notion of a set *enumerable with respect to* α (or α-*enumerable*) in any of the following equivalent ways: as the domain of an α-computable function, as the range of an α-computable function, as the projection of an α-decidable (decidable relative to α) set, and so on. But there also is a more direct description of the class of α-enumerable sets.

Let E be an arbitrary set of pairs of the form $\langle x, t \rangle$, where x is a number and t is a pattern. We take a total function α and pick out from the set E the pairs whose second components are parts of α; the first components of these pairs form a set that will be denoted by $E[\![\alpha]\!]$.

Theorem 46. *A set X is α-enumerable if and only if $X = E[\![\alpha]\!]$ for some enumerable set E. (Notice that in this case no special conditions like consistency of the set are needed.)*

Proof. Let X be the domain of a function f computable relative to α. Then $f = M[\alpha]$ for some enumerable consistent set M. Let us delete the second component of each triple from M. We obtain an enumerable set of pairs; denote it by E. It is easy to see that $E[\![\alpha]\!]$ is the domain of the function $M[\alpha] = f$, and so $E[\![\alpha]\!] = X$.

Conversely, suppose that $X = E[\![\alpha]\!]$ for some function α. Then consider the set M obtained by inserting 0 between the components of each pair in E. Clearly, the set M is consistent and $M[\alpha]$ is the function defined on $X = E[\![\alpha]\!]$ that takes only zero values. $\qquad\square$

In the ordinary (nonrelativized) theory of algorithms, an important part is played by the universal function theorem. It remains true after relativization.

Theorem 47. *Let α be a total function. There exists a binary α-computable function universal for the class of unary α-computable functions.*

Proof. As well as in other cases, we can reproduce the proof of the corresponding nonrelativized theorem almost unchanged. Let us fix any programming language (this time involving calls to external procedures) and assign numbers to all programs with calls to the external procedure α. Now we can take as universal function the function

$$U_\alpha(i, x) = (\text{result of application of the } i\text{th program to } x).$$

We have used the subscript to emphasize that the function U_α depends on α. However, the text of the program that computes U_α does not depend on α (although, naturally, it contains calls to α).

It is instructive to look at another proof, based on the definition of computability in terms of consistent enumerable sets.

Consider a universal enumerable set Z of quadruples of the form $\langle n, x, y, t \rangle$, where n, x, and y are numbers, and t is a pattern. The term "universal" means here that any enumerable set of triples coincides with one of the sections Z_n for a suitable n.

There can be both consistent and inconsistent sets among Z_n. We want to forcedly correct the inconsistent sections leaving the consistent ones intact. In other words, we want to construct a new enumerable set Z' with the following properties: first, all sections Z'_n must be consistent; second, if a certain section Z_n is consistent, then it should not change ($Z'_n = Z_n$).

The construction is simple: we enumerate Z and skip (do not put into Z') the elements that make a certain section inconsistent when added to it. Thus we get an enumerable set Z' universal for the class of consistent enumerable sets.

Now it is easy to specify a consistent set W that yields a universal α-computable function. Namely, the set W consists of the triples $\langle \langle n, x \rangle, y, t \rangle$ (their first components are pairs, because a universal function depends on two arguments) such that $\langle n, x, y, t \rangle \in Z'$. It can readily be seen that W is a consistent set. For a given function α, this consistent set specifies a certain α-computable binary function U_α; its nth section is $Z'_n[\alpha]$, where Z'_n is the nth section of the set Z'. Therefore, any α-computable function is a section of the function U_α, completing the proof.   □

Of course, the notion of Gödel universal function has a counterpart in the relativized theory of algorithms as well: a binary α-computable function is called a *Gödel universal function for the class of unary α-computable functions* if it is α-computable, universal for the class of unary α-computable functions, and for any binary α-computable function V there exists a unary total α-computable function s ("convertor") such that $V(n, x) = U(s(n), x)$ for all n and x.

The standard proof (p. 20, Theorem 15) shows that Gödel universal functions for the class of α-computable functions do exist. Furthermore, we can notice that the function s constructed in the proof is not only α-computable, but also computable (one version of the proof uses a function s of the form $x \mapsto [n, x]$, where the brackets denote a fixed computable numbering of pairs and n is a fixed integer).

It is convenient to understand by numbers of α-computable functions their numbers under such "strongly Gödel" numberings. In particular, if we list all programs with oracle calls (in some natural ordering) and then assign to each program its number in the list, we get a "strongly Gödel" numbering.

Sometimes, when it goes about the relativized theory of algorithms, the following metaphor is used. Let A be an undecidable set. It may happen that there is an extraterrestrial civilization to which the set A seems decidable. They can decide whether a number x belongs to the set A directly, just by looking at it; this test is an elementary operation in their programs (like, e.g., the comparison of two numbers is an elementary operation in our programs). Then their entire theory of algorithms will automatically be relativized with respect to A, but they will never notice this, and, therefore, having read our book up to this section (exclusively), will agree with all our theorems. Furthermore, they can also read this section about relativization as well, but whatever is B-computable to them, will be A-B-computable for us (computable with the help of two oracles, for A and B).

However, you should not take this metaphor too seriously.

4. 0′-computations

In this section we consider computability with respect to an m-complete enumerable set. Any two of these sets can be m-reduced to each other, and a fortiori they can be T-reduced to each other. Therefore, if some function is computable relative to one of them, then it is computable relative to the other as well. Such functions are called $\mathbf{0}'$-*computable*.

Recalling that the set of pairs $\{\langle p, x \rangle \mid$ program p halts on the input $x\}$ is an m-complete enumerable set, we can say that $\mathbf{0}'$-computable functions are computed by machines supplied with a special oracle that solves the halting problem: we send a program and an input to this oracle and it answers whether the program halts on this input or not. (The program sent for examination is a normal one, without any calls to an oracle.)

Clearly, any enumerable set is $\mathbf{0}'$-decidable, since it is m-reducible to an m-complete enumerable set. (The converse is obviously not true: the complement of an enumerable undecidable set is also $\mathbf{0}'$-decidable, but not enumerable.)

The following theorem gives a simple description of the class of $\mathbf{0}'$-computable functions.

Theorem 48. (a) *Let T be a total computable function of two natural arguments. Take the limit in the second argument and consider the function t thus obtained:*

$$t: x \mapsto \lim_{n \to \infty} T(x, n).$$

(This function may not be total, since for some x the limit may not exist.) The function t is $\mathbf{0}'$-computable.

(b) *Any $\mathbf{0}'$-computable function t can be so obtained from some total computable function T.*

Proof. (a) Let T be a binary total computable function. We will say that a pair $\langle x, n \rangle$ is stable if $T(x, n) = T(x, m)$ for the given x and for all $m > n$. Notice that the set of unstable pairs is enumerable (having found two pairs $\langle x, n \rangle$ and $\langle x, m \rangle$ with $n < m$ and $T(x, n) \neq T(x, m)$, we include the pair $\langle x, n \rangle$ in the enumeration of all unstable pairs). Therefore, the set of unstable pairs is $\mathbf{0}'$-decidable. In other words, $\mathbf{0}'$-algorithm can check stability of any pair.

Now consider the following $\mathbf{0}'$-algorithm computing the limit function t. Having received an input x, we consider the pairs $\langle x, 0 \rangle, \langle x, 1 \rangle, \ldots$ and for each of them check whether it is stable. Once a stable pair $\langle x, n \rangle$ is detected, the value $T(x, n)$ is sent to the output. Obviously, this $\mathbf{0}'$-algorithm computes the function t.

(b) Now let us prove the converse statement. Let t be a partial $\mathbf{0}'$-computable unary function. We must construct a computable (without an oracle) binary total function T such that

$$t(x) = \lim_{n \to \infty} T(x, n)$$

for all x (this means that both sides of this equation are defined simultaneously and coincide when defined). For a while, let us allow the function T to take some additional special value, which will be denoted by a star. This value is not allowed to be the limit value, and $\lim_{n \to \infty} T(x, n) = a$ means that for all sufficiently large n the value $T(x, n)$ is equal to a (and not equal to $\star$).

This allowance is in fact inessential: we can always replace each star in a sequence by two different successive terms (no matter which); then the new sequence will have the same limit (or have no limit, as the original one).

Now let us define the function T. By assumption, the function t is computed by a certain program p that has access to the characteristic function of some enumerable set K. Denote by K_n the finite subset of K consisting of the elements of K that appeared during the first n steps of its enumeration. To compute the value $T(x, n)$, we run n steps of the program p with the set K replaced by its finite approximation K_n. If we get no answer from the program during these n steps (which can happen for various reasons: the time may be too short for p to produce a result, the difference between K_n and K may come into play, or t can simply be undefined at x), then we set $T(x, n) = \star$. And if the program produces some answer in n steps, then this answer is set to be the value of $T(x, n)$ (with the only exception explained later).

Let us try to prove that $t(x) = \lim_{n \to \infty} T(x, n)$. Suppose that $t(x)$ is equal to some a. Then, after several steps, the execution of p (with the regular oracle K) terminates, and the program returns the answer a. In this computation, only a finite number of questions to the oracle are asked. Therefore, for a sufficiently large n, the answers to these questions will remain the same if K-oracle is replaced by K_n-oracle. By taking, if necessary, a still greater value of n (so as to exceed the total computation time of the program p on the input x), we can ensure that for this and all greater values, $T(x, n)$ is equal to a.

We have yet to show that if the limit exists and equals a, then $t(x) = a$. But here an additional obstacle awaits us: It is possible that the program p does not halt using the original K-oracle, but the computation with the K_n-oracle terminates for each n (due to the differences between K_n and K), accidentally returning the same answer every time. Then $t(x)$ is undefined, but $\lim_{n\to\infty} T'(x,n)$ exists, and this is bad for us.

To overcome this difficulty, let us modify the definition of the function T. Namely, let us agree that if the oracle query logs created during the computation of $T(x,n)$ and $T(x,n-1)$ were different (i.e., different questions were asked or different answers to the same questions were received), then $T(x,n) = \star$. This will not corrupt our previous reasoning, since, for large n, the questions involved in the computation of $T(x,n)$ and the answers to them are the same as in the original K-computation.

But we can now be sure that if the sequence $T(x,0), T(x,1), \ldots$ has a limit, then $t(x)$ is defined. Indeed, if the limit exists, then the sequence contains finitely many stars. Hence for all sufficiently large n, the oracle is asked the same questions and gives the same answers. This means that these answers are correct, because K_n tends to K. Therefore, the original computation of $p(x)$ using the K-oracle also terminates (with the same result).   □

Problem 60. The definition of computable real numbers given in Problem 14 (p. 9) can be relativized with respect to any set A. Show that a number α is $0'$-computable if and only if α is the limit of a computable sequence of rational numbers.

5. Incomparable sets

The definition of Turing reducibility (recall that A is Turing reducible to B if the set A is decidable with the help of an oracle for B) can be viewed as a method to compare decidability problems for various sets in their "difficulty." (If $A \leq_T B$, then the decidability problem for the set A is, in a sense, simpler than that for the set B.)

A lot of natural questions concerning this classification arise. For instance, we may want to know if there exists the most difficult decidability problem in the world, that is, a set A such that $B \leq_T A$ for any set B. It is readily seen that the answer is negative:

the A-relativized world has its own undecidable sets (and even A-enumerable A-undecidable sets), since it obeys usual laws of the theory of algorithms. (One can also notice that since there are countably many programs, the family of all A-decidable sets is countable for any set A.)

Another, less trivial, question is: Are any two sets comparable? The following theorem proved by Kleene and Post shows that here the answer is also negative.

Theorem 49. *There exist two sets A and B such that $A \not\leq_T B$ and $B \not\leq_T A$. We can choose these sets to be $\mathbf{0}'$-decidable.*

Proof. The sets A and B must satisfy the following requirements: no B-oracle program (i.e., a program supplied with a B-oracle) decides the set A, as well as no A-oracle program decides the set B.

Thus we have countably many requirements (since there are countably many programs). We will treat them in turn, each once: having ensured that a requirement is fulfilled, we will never consider it again. Each step will fix the behavior of the sets A and B on some initial segments of $\mathbb{N}$; after that we can be sure that all the requirements considered up to this step are satisfied. At the next step the initial segments where A and B are fixed increase; in the limit, we will obtain two sets, A and B, satisfying all the conditions. The entire construction will be $\mathbf{0}'$-computable, so the resulting sets will be $\mathbf{0}'$-decidable.

Here are the details. Any function defined on a (finite) initial segment of $\mathbb{N}$ that takes the values 0 and 1 will be called a *fragment*. We say that a set A *complies* with a fragment $a \colon \{0, \ldots, m\} \to \{0, 1\}$ if the characteristic function of A extends a. In other words, compliance with a means a certain behavior of a set on natural numbers that do not exceed m.

If a fragment a_2 extends a fragment a_1 (i.e., a_2 is defined on a greater segment and coincides with a_1 on the domain of a_1), then, obviously, the compliance with a_2 is more restrictive for a set.

Lemma. *Let a and b be two fragments, and let p be a program that includes calls to an external procedure. Then there exist extensions a' and b' of these fragments with the following property: for any sets A*

and B complying with a' and b', the program p, using the B-oracle, does not decide the set A.

Assuming the Lemma has been proved, we can consider all programs one by one and ensure that none of them decides A using the B-oracle. In so doing, we can alternate A and B as the Lemma is applied, and thus simultaneously ensure that no program decides B using the A-oracle.

(Remark. Obviously, we can assume that a' and b' are longer than a and b: otherwise, they can be extended artificially. Then in the limit we will obtain infinite sequences that are characteristic functions of the two desired sets. However, this is not necessary: if we incidentally obtain finite fragments in the limit, we can take any two arbitrary sets complying with them.)

So, to complete the construction of the sets A and B, it remains to prove the Lemma. (We will return to the issue of $\mathbf{0}'$-computability later.)

The statement of the Lemma is asymmetric with respect to the sets A and B; therefore, the reasoning will also be asymmetric. Let us fix a number x outside the domain of the fragment a. The argument depends on the answer to the question: Does there exist a set B complying with the fragment b such that the program p with the B-oracle returns one of the answers "yes" or "no" for input x?

If such a set does not exist, there is nothing at all to worry about: the statement of the Lemma will be true simply for $a' = a$ and $b' = b$.

Now suppose that such a set B exists. Let us run and trace the program p on the input x for this set B. Before returning an answer, the program p can make a finite number of calls to the characteristic function of the set B. Let us take a fragment b' with which B complies and long enough to cover all the elements involved in the program calls. Then for any set complying with b', the program p will return the same answer as for the set B. It remains to ensure that this answer is wrong. This can be done by including x into the domain of a' and choosing $a'(x)$ to contradict this answer. The proof of the Lemma is complete.

Finally, it remains to prove the statement of the theorem concerning the $0'$-computability; to this end, we must ascertain that the construction of a' and b' in the proof of the Lemma can be made $0'$-algorithmic. The key point here is the answer to the question posed in the proof of the Lemma. Of course, it is impossible to test all sets B complying with the fragment b. But we do not have to really do this: it suffices simply to examine all the paths that the program p can follow. When it asks the question about a number outside b, we must consider two possibilities. Thus a tree of all possible computation paths emerges, and we want to know whether we get the answer "yes" or "no" at least on one of the branches. This question can be reformulated into the question of whether a certain program halts (namely, the program that examines all branches in parallel and terminates once the answer "yes" or "no" is given on one of them), and the $0'$-oracle can provide an answer to this question.

This remark completes the proof of the theorem. $\square$

A much more difficult question is: Do there exist *enumerable* Turing incomparable sets (rather than $0'$-decidable Turing incomparable sets)? This problem (called the *Post problem*) was solved independently by the American mathematician Richard Friedberg and the Russian mathematician Al'bert Abramovich Muchnik in the 1950s; it is interesting that their constructions of enumerable incomparable sets used the same approach, called the "priority method". This construction is explained in the next two sections.

6. Friedberg–Muchnik Theorem: The general scheme of construction

Theorem 50. *There exist Turing incomparable enumerable sets.*

We give the proof of this theorem as an example of the technique used in the theory of computable functions. However, it should be mentioned that in the 1960s and 1970s much more sophisticated methods appeared, and now this proof looks relatively simple.

Proof. We want to construct two enumerable sets, none of which is Turing reducible to the other. We will construct them step by step;

at each step, only a finite part of the sets to be created will be known. It will be convenient to use the following terminology.

Any pair $\langle A, B \rangle$ of finite sets of natural numbers will be called an *element*. We will say that an element $\langle A', B' \rangle$ *extends* an element $\langle A, B \rangle$ if $A \subset A'$ and $B \subset B'$. We will construct a computable sequence of elements; each element extends the previous one, and in the limit (i.e., in the union) they form the desired enumerable incomparable sets.

A quadruple of finite sets $\langle A^+, A^-, B^+, B^- \rangle$ in which A^+ is disjoint with A^- and B^+ is disjoint with B^- will be called a *directive*. The word "directive" is derived from the fact that these quadruples will direct the construction of elements: A^+ are the numbers that must belong to A, and A^- are the numbers that must not belong to A; similarly for B. Formally, we say that an element $\langle A, B \rangle$ *complies* with a directive $\langle A^+, A^-, B^+, B^- \rangle$ if $A^+ \subset A$, $A^- \cap A = \varnothing$, $B^+ \subset B$, and $B^- \cap B = \varnothing$. A directive u_2 is said to be *stronger* than a directive u_1 if any element complying with u_2 complies with u_1 also (that is, all the components of the stronger directive are larger). Note that according to this definition, any directive is stronger than itself.

Let $\alpha(X, Y)$ be an arbitrary condition on a pair of sets $X, Y \subset \mathbb{N}$. Any such condition specifies a game for two players, the "director" D and the operator O. The game proceeds as follows: in the beginning O presents to D a directive u_0 and an element e_0 complying with u_0. They will be called the initial directive and initial element. (Eventually, the construction will involve several directors, and newly appointed directors will receive their initial directives and elements from their predecessors; but this is to be discussed later.) Coming back to the game, D replies to O with some directive u_1; after that, O chooses an element e_1 complying with this directive; then D issues some directive u_2, O chooses e_2, and so on (the game is infinite).

The rules of the game are as follows:

- Each element chosen by O must extend the previous one (and so all of them extend the initial element); it must also comply with the last directive of D (but D's earlier directives are not important).

- Each directive of D must be stronger than the initial directive (however, we do not require it to be stronger than D's earlier directives!).

- If D issues a directive that causes a stalemate (that is, there are no elements complying with this directive and extending the previous element), then the game terminates and D loses.

- If the game goes on indefinitely, then D is considered the winner under two conditions. The first of them is that starting from a certain move, D's directives stop changing.

- Finally, the second condition is that the limit sets X and Y should satisfy the above-mentioned condition $\alpha(X, Y)$. (If the ith element e_i is $\langle X_i, Y_i \rangle$, then X and Y are the unions of increasing chains of sets $X_0 \subset X_1 \subset \cdots$ and $Y_0 \subset Y_1 \subset \cdots$.)

A condition $\alpha(X, Y)$ is said to be *winning* if there exists a computable winning strategy in the corresponding game. Our further plan is as follows. We will show that for any program p with calls to external procedure, the condition $\alpha_p(X, Y) =$ "p with Y-oracle does not decide X" is a winning one. (Much of the reasoning repeats the proof of the Kleene–Post Theorem, but in a more intricate way.) Moreover, we will see that the strategy computably depends on p.

On the other hand, we will show how to construct a pair of enumerable sets that satisfies a countable family of winning conditions. It is this last argument that makes use of the "priority" idea: we will have one operator and countably many directors with different priorities assigned to them.

7. Friedberg–Muchnik Theorem: Winning conditions

So, let us fix a program p that the director D wants to prevent from deciding the set X relative to Y. What must D do for that? (We will describe the course of events from D's viewpoint.)

In the beginning, D receives from O a directive and an element complying with it. All subsequent directives must be stronger than the initial one: we (the director) should always prescribe that certain (initial) numbers be included in X and Y, whereas certain other numbers should not (the sets of numbers of both kinds are finite). Besides, an initial element (pair of sets) is given. As time goes by, O increases these sets at his discretion; the only way for us to influence this process is to issue directives.

So what do we do? At the first step we choose a number x outside the initial set X and not mentioned in the initial directive. In our first directive we will require that this x not be included in X, that is, we will add x to the second component of the initial directive, once denoted by A^-. (If we wished the set Y to be undecidable by the program with X-oracle, we would act symmetrically and add the number to the fourth component, B^-.)

If we keep replicating the first directive on and on, we will ensure that the number x does not belong to the limit set X. But if at a certain instant we change our mind and choose to include x in X, it will suffice merely to take x away from A^- and add it to A^+, which will not stalemate the game. (Notice that the new directive will not be stronger than its predecessor, but will remain stronger than the initial one, which is required by the game rules.)

Our winning strategy chooses such an x, forms the first directive, and keeps repeating it until we have a reason to wake up and change our mind. Such a reason can be as follows:

At the nth move of the game, we simulate n steps of the program p on the input x. (Recall that p is the program we want to prevent from Y-deciding the set X.) In so doing, every time the program invokes the external procedure for Y, we answer using the current state of Y (that is, according to the last element proposed by the operator). Imagine that after the n steps the program p returns a certain result. Then we wake up and examine the computation to spot the numbers whose membership or nonmembership in the set Y was used; the results of the examination are registered in our next and all subsequent directives (which will never change any longer). This will ensure that p will give the same answer with the limit set Y as it gives now (i.e., for the current state of Y). On the other hand, we can place x in the set X or outside of it at will. (If we do nothing and leave things as they are, x will not belong to X; to move it into X, we simply move it to the positive component of the directive; see above.) Our choice will be to make the answer of the program p wrong.

Let us show that this strategy is winning indeed. There are two possibilities. If we have woken up at a certain move, then, by construction, the answer returned by the program p on the input x is wrong. Otherwise, if we never wake up, p (with the limit value of the oracle Y) does not return any answer on x at all. Why? Because any answer it may return depends on a finite number of questions to the oracle and requires a finite number of program steps; so, after the game has been played sufficiently long (enough for the oracle to include all the necessary numbers and for the program to complete computation), we would have to wake up.

The computability of our strategy is obvious. It only remains to explain why the number of different directives will be finite. But this obviously follows from the construction. In fact, there can be at most two: exactly two, if we have ever woke up to fall asleep forever; otherwise, just one.

Note that the winning strategy in the game with the program p can be constructed effectively (by an algorithm) when p is given. This observation plays an important role in the next section.

8. Friedberg–Muchnik Theorem: The priority method

Now we forget about the specific nature of elements and directives and show that if there is a sequence of winning conditions $\alpha_1, \alpha_2, \ldots$ such that the corresponding winning strategies for D computably depend on i, then there is a pair of enumerable sets satisfying all the conditions.

To this end, imagine ourselves being an operator under a sequence of directors with decreasing priorities (the first director is the most important, the second is less important, and so on). All the directors are responsible for their own conditions and have their own winning strategies. We begin the game with the first director by executing her directives. When they start to repeat, we assume that the first director will not change them any more and pass the current element and current directive as the initial data to the second director to start the game with the latter. (All directives of the second director will be stronger than the temporarily stabilized directives of the first. Therefore, we do not violate the rules of the game played with the first director, and at the same time obey the second director, unless the first one starts changing her directives.) When the directives of the second director also stop changing, we can link up the third one, and so on.

What happens if one of the directors unexpectedly changes her directive? In this case we follow this new directive paying no attention to what the directors of lower priority say, apologize to them, and explain that we have invited them too early. But later we gradually invite them back according to the same scheme.

Let us show that each director will sooner or later have a chance to direct permanently. Indeed, the first one does not know anything at all about the others, because her directives are the most important and are always obeyed. Therefore, a moment will come when she ceases to issue new directives. The incarnation of the second director launched after this moment will rule unhindered, because only a change of the first director's directives could interrupt the second one's directorship. Therefore, directives of the second director also stabilize, and so on. In so doing, all directors will secure the fulfillment of their respective conditions.

Now we see why the initial elements and conditions were important in the definition of our game: a true Director must be able to achieve her objectives irrespective of the initial situation (who knows what the predecessors could do to the business!).

The computability of all strategies of all directors (plus the computable dependence on directors' numbers) guarantees the computability of the above-described process; thus we obtain enumerable sets satisfying all our conditions.

This argument completes the proof of the Friedberg–Muchnik Theorem.   □

Problem 61. Show that there are countably many enumerable sets, no two of which are Turing comparable.

Chapter 8

Arithmetical Hierarchy

1. Classes Σ_n and Π_n

As we have already said, enumerable sets can be equivalently defined as projections of decidable sets: a set $A \subset \mathbb{N}$ is enumerable if and only if there exists a decidable set $B \subset \mathbb{N} \times \mathbb{N}$ such that A is a projection of B. Identifying sets and properties ($=$ predicates), we can say that a property $A(x)$ of natural numbers is enumerable if and only if it can be represented in the form

$$A(x) \Leftrightarrow \exists y B(x, y),$$

where $B(x, y)$ is a certain decidable property.

(In this section we assume that the reader is familiar with the basic logical notation: the quantifier $\exists x$ is read as "there exists x", the quantifier $\forall x$ is read as "for all x", the symbol $\wedge$ read as "and" is called *conjunction*, the symbol $\vee$ read as "or" is called *disjunction*, the symbol $\neg$ read as "it is not true that" is called *negation*. As before, the symbol $\Leftrightarrow$ denotes equivalence.)

A natural question arises: What can be said about other combinations of quantifiers? For instance, what properties are representable in the form

$$A(x) \Leftrightarrow \exists y \exists z C(x, y, z),$$

where C is a decidable property of triples of natural numbers? It is readily seen that these are again enumerable sets. Indeed, two consecutive quantifiers of the same type can be replaced by one using a computable numbering of pairs (which is denoted by brackets): a property C' such that $C'(x, [y, z]) \Leftrightarrow C(x, y, z)$ is also decidable, and $A(x) \Leftrightarrow \exists w C'(x, w)$.

Another question: What properties can be represented in the form

$$A(x) \Leftrightarrow \forall y B(x, y),$$

where $B(x, y)$ is a decidable property? The answer is: the properties with enumerable negations (also known as *coenumerable*). Indeed,

$$\neg A(x) \Leftrightarrow \neg \forall y B(x, y) \Leftrightarrow \exists y (\neg B(x, y));$$

it remains to notice that decidability is preserved under negation.

Let us give a general definition. A property A belongs to the *class Σ_n* if it can be represented in the form

$$A(x) \Leftrightarrow \exists y_1 \forall y_2 \exists y_3 \ldots B(x, y_1, y_2, \ldots, y_n)$$

(with n alternating quantifiers in the right-hand side), where B is a decidable property. If the n alternating quantifiers in the right-hand side start with the universal quantifier $\forall$, then we obtain the definition of the *class Π_n*.

The following two properties, in essence, have already been proved.

Theorem 51. (a) *The class Σ_n $[\Pi_n]$ does not change if we allow groups of quantifiers of the same type ($\forall$ or $\exists$) instead of a single $\forall$- or $\exists$-quantifier.* (b) *If a predicate belongs to Σ_n, its negation belongs to Π_n and vice versa.*

(The statement (a) implies, for example, that the predicate

$$\exists y \, \exists z \, \forall u \, \forall v \, \forall w \, \exists t \, A(x, y, z, u, v, w, t)$$

with decidable A belongs to Σ_3 since there are three groups starting with the $\exists$-group.)

Proof. To prove the first statement, it suffices to combine neighboring quantifiers of the same type into one quantifier using the numbering of pairs. To prove the second statement, we use the laws $\neg \forall x\, A \Leftrightarrow \exists x\, \neg A$ and $\neg \exists x\, A \Leftrightarrow \forall x\, \neg A$ and recall that the negation of a decidable predicate is decidable. □

We spoke about properties (predicates); in terms of sets, the definition of the class Σ_n takes the following form: sets of the class Σ_n are obtained from decidable sets by a sequence of operations "projection–complement–projection–complement–...– projection" with exactly n projections. Each projection decreases the dimension of the set (the number of arguments of the corresponding property) by one, so we must start with decidable subsets of $\mathbb{N}^{n+1}$.

Theorem 52. *The intersection and union of two sets of the class Σ_n belong to Σ_n. The intersection and union of two sets of the class Π_n belong to Π_n.*

Proof. We have to prove that the conjunction and disjunction of any two properties of the class Σ_n belong to this class again (and similarly for Π_n). For instance, suppose that

$$A(x) \Leftrightarrow \exists y \forall z B(x, y, z),$$
$$C(x) \Leftrightarrow \exists u \forall v D(x, u, v).$$

Then

$$A(x) \wedge C(x) \Leftrightarrow \exists y \exists u \forall z \forall v [B(x, y, z) \wedge D(x, u, v)];$$

the property written in the brackets is decidable, and it only remains to combine pairs of quantifiers as described above. The classes Σ_n and Π_n for an arbitrary n can be treated similarly. □

The classes Σ_n and Π_n were defined for sets of natural numbers; in a similar way this can be done for sets of pairs or triples of natural numbers and, in general, for any "constructive objects". Notice that the projection of a set of pairs from the class Σ_n also belongs to Σ_n (since two existential quantifiers can be combined into one).

By adding dummy quantifiers, it is easy to show that each of the two classes Σ_n and Π_n is contained in each of the classes Σ_{n+1}

and Π_{n+1}. We can write this as follows:

$$\Sigma_n \cup \Pi_n \subset \Sigma_{n+1} \cap \Pi_{n+1}.$$

Theorem 53. *The classes Σ_n and Π_n are "hereditary downward" with respect to m-reducibility in the following sense: if $A \leq_m B$ and $B \in \Sigma_n$ $[B \in \Pi_n]$, then $A \in \Sigma_n$ $[A \in \Pi_n]$.*

Proof. Suppose that A is reduced to B by a total computable function f, that is, $x \in A \Leftrightarrow f(x) \in B$. Suppose, for example, that B belongs to Σ_3:

$$x \in B \Leftrightarrow \exists y \forall z \exists u R(x, y, z, u),$$

where R is some decidable property. Then

$$x \in A \Leftrightarrow f(x) \in B \Leftrightarrow \exists y \forall z \exists u R(f(x), y, z, u),$$

and it remains to notice that $R(f(x), y, z, u)$ (as a property of a quadruple $\langle x, y, z, u \rangle$) is decidable.   □

Problem 62. Prove that if a set A belongs to the class Σ_n, then the set $A \times A$ also belongs to this class.

Problem 63. Prove that if sets A and B belong to the class Σ_n, then their set difference $A \setminus B$ belongs to the class $\Sigma_{n+1} \cap \Pi_{n+1}$.

2. Universal sets in Σ_n and Π_n

We have not yet shown that the classes Σ_n, as well as Π_n, are distinct for different n. To prove this, we will find in each of these classes a universal set (for the corresponding class) and show that it does not belong to junior classes.

Theorem 54. *For any n, the class Σ_n contains a set universal for all sets of this class. (The complement of this set is universal in the class Π_n.)*

By a universal set of the class Σ_n (briefly, Σ_n-universal set) we mean a set of pairs of natural numbers that belongs to the class Σ_n such that any set of natural numbers in Σ_n is a section of this set of pairs.

Proof. Notice that Σ_1 is the class of enumerable sets. The existence of a universal set for this class has already been discussed. Universal sets for higher classes of the hierarchy will be constructed using this set. (We have to start from the first level, because at the "zeroth level" there are no universal decidable sets.)

By the definition, Π_2-properties are of the form $S(x) \Leftrightarrow \forall y \exists z R(x, y, z)$, where R is a decidable property. But they can also be defined equivalently as properties of the form $S(x) \Leftrightarrow \forall y P(x, y)$, where P is an enumerable property. Now it is clear how to construct a universal set of the class Π_2. Let $U(n, x, y)$ be a universal enumerable property. Then any enumerable property of pairs of natural numbers can be obtained from $U(n, x, y)$ by fixing a suitable n. Hence any Π_2-property of natural numbers can be obtained from the property $T(n, x) = \forall y U(n, x, y)$ in the same way, by fixing n. On the other hand, the property T itself belongs to the class Π_2.

The complement of a universal Π_2-set is obviously a universal Σ_2-set.

This reasoning applies to Σ_3- and Π_3-sets, as well, with one amendment: here it is better to start with Σ_3-sets, so as to have the existential quantifier, which specifies an enumerable set, in the innermost position. Σ_n- and Π_n-sets are considered similarly. □

Theorem 55. *Universal Σ_n-sets do not belong to the class Π_n. Similarly, universal Π_n-sets do not belong to the class Σ_n.*

Proof. Consider a universal Σ_n-property $T(m, x)$. By definition, this means that all Σ_n-properties occur among its sections (obtained by fixing m). Suppose that T belongs to the class Π_n. Then its diagonal, the property $D(x) = T(x, x)$, also belongs to the class Π_n (for example, because $D \leq_m T$), and the negation of D, the property $\neg D(x)$, belongs to the class Σ_n. But this is impossible, since $\neg D$ is not a section of the property T (it differs from the mth section at the point m), whereas T is universal. □

In particular, it follows from this theorem that any of the classes Σ_n and Π_n is a proper subset of any of the classes Σ_{n+1}

and Π_{n+1}. (Soon we will see that even the union $\Sigma_n \cup \Pi_n$ is a proper subset of the intersection $\Sigma_{n+1} \cap \Pi_{n+1}$.)

3. The jump operation

We want to show that the class Σ_n coincides with the class of all A-enumerable sets for a certain set A (depending on n). To introduce this set, we need a construction known as the jump operation.

Let X be an arbitrary set. Consider the class of all X-enumerable subsets of $\mathbb{N}$ and a universal X-enumerable set for this class. This set is m-complete in the class of X-enumerable sets in the sense that all other X-enumerable sets are m-reducible to it. The reducing function, as we have seen, is of the form $x \mapsto [n, x]$ (and is computable without any oracle, as the definition of m-reducibility requires). We will denote by X' any m-complete set in the class of X-enumerable sets. Such a set is defined uniquely up to m-equivalence.

More formally, we say that the sets P and Q are *m-equivalent* if $P \leq_m Q$ and $Q \leq_m P$. (It can be readily seen that this is indeed an equivalence relation.) A class of equivalent sets is called an *m-degree*. Now we can say that for each set X we have defined a certain m-degree X'.

In a similar way, *T-degrees* (also known as *Turing degrees* or *degrees of undecidability*) are defined as classes of T-equivalent sets; sets P and Q are called *T-equivalent*, or *Turing equivalent*, if $P \leq_T Q$ and $Q \leq_T P$, that is, if either set is decidable relative to the other. If the sets P and Q are Turing equivalent, then the class of P-computable functions coincides with the class of Q-computable functions (and the class of P-enumerable sets coincides with the class of Q-enumerable sets). Using the notion of T-degrees, we can say that the m-degree X' is determined by the T-degree of the set X, and thus we have a mapping of the set of all T-degrees to the set of all m-degrees. This mapping is called the *jump operation*; the set (more exactly, m-degree) X' is called the *jump* of the set (more exactly, T-degree) X.

Problem 64. Can this mapping take different T-degrees into the same m-degree?

Problem 65. Prove that any two sets m-complete in the class Σ_n are computably isomorphic (differ by a computable permutation).

Problem 66. Show that for any enumerable set A, one can find a real number α such that the set of all rational numbers less than α is enumerable and Turing equivalent to the set A.

The jump operation is usually considered as an operation on T-degrees by setting its result equal to the T-degree containing X' (this is quite legal, since the T-classification is more coarse).

In the sequel, we use the following T-degrees: $\mathbf{0}$ (the degree containing all decidable sets), $\mathbf{0}'$ (its jump, the degree of m-complete enumerable sets; it has already been considered), then $\mathbf{0}''$ (the jump of the degree $\mathbf{0}'$), $\mathbf{0}'''$, and so on; in general, $\mathbf{0}^{(n+1)} = (\mathbf{0}^{(n)})'$.

Theorem 56. *For any $n \geq 1$, the class Σ_n coincides with the class of all $\mathbf{0}^{(n-1)}$-enumerable sets.*

(So far we know this for $n = 1$.)

Proof. First, we will prove that all Σ_n-sets are enumerable with respect to $\mathbf{0}^{(n-1)}$. This is done by induction over n. For $n = 1$, we already know this. Now consider an arbitrary set X from Σ_2. By the definition of Σ_2,

$$x \in X \Leftrightarrow \exists y \forall z R(x, y, z),$$

where R is a decidable predicate. The predicate $\forall z R(x, y, z)$ has enumerable negation. This negation is decidable with respect to $\mathbf{0}'$, since it is m-reducible to an m-complete enumerable set. Hence the predicate $\forall z R(x, y, z)$ itself is decidable with respect to $\mathbf{0}'$. Therefore, its projection, the set X, is enumerable with respect to $\mathbf{0}'$.

For other values of n, the argument is similar. If X belongs to Σ_3, then

$$x \in X \Leftrightarrow \exists y R(x, y),$$

where R belongs to Π_2. The negation of R belongs to Σ_2, so it is $\mathbf{0}'$-enumerable (by the induction assumption), so it is $\mathbf{0}''$-decidable, so R itself is $\mathbf{0}''$-decidable as well, and so the projection of R is $\mathbf{0}''$-enumerable.

This completes the first half of the proof.

For the second half, we must introduce one more property of the classes Σ_n and Π_n. Consider some computable numbering of all finite sets of natural numbers. Denote by D_x the finite set whose number is x. For an arbitrary set A, consider the set Subset(A) of all finite subsets of A, more exactly, the set of all their numbers:

$$x \in \text{Subset}(A) \Leftrightarrow D_x \subset A.$$

Lemma 1. *If a set A belongs to the class Σ_n [or Π_n], then the set* Subset(A) *also belongs to the class Σ_n [Π_n, respectively].*

(The statement of this lemma generalizes the statement about the set $A \times A$ formulated in Problem 62: now we consider arbitrary tuples instead of pairs.)

Proof of Lemma 1. Consider, for instance, a set A of the class Σ_3:

$$x \in A \Leftrightarrow \exists y \forall z \exists t R(x, y, z, t),$$

where R is a decidable property. Then $\{x_1, \ldots, x_n\} \subset A$ is equivalent to

$$\exists\langle y_1, \ldots, y_n\rangle \forall\langle z_1, \ldots, z_n\rangle \exists\langle t_1, \ldots, t_n\rangle [R(x_1, y_1, z_1, t_1) \wedge \ldots$$
$$\ldots \wedge R(x_n, y_n, z_n, t_n)].$$

This formula involves quantifiers over tuples of natural numbers (of variable length), but we can replace tuples by their numbers in some computable numbering. Moreover, since the inner quantifier-free formula corresponds to a decidable set, the entire formula is a Σ_3-predicate.

In this argument we made no difference between tuples and finite sets; this is safe since the transition from the number of a set to the number of some tuple consisting of all its elements is computable, so this is no problem.

Lemma 1 is proved.   □

Problem 67. Prove that if A belongs to the class Σ_n [Π_n], then the set Intersect(A) of the numbers of finite sets having nonempty intersections with A also belongs to the class Σ_n [Π_n].

Problem 68. Suppose that a property $R(x, y)$ of pairs of natural numbers belongs to the class Σ_n. Show that the property

$$S(x) = (\forall y \leq x) \, R(x, y)$$

belongs to Σ_n. (The bounded quantifier $(\forall y \leq x)$ is read as "for all y not exceeding x".)

Passing to the complements, we immediately obtain the following corollary to Lemma 1:

Lemma 2. *If a set A belongs to the class Σ_n $[\Pi_n]$, then the set $\mathrm{Disjoint}(A)$ consisting of the numbers of finite sets disjoint with A belongs to Π_n $[\Sigma_n]$.*

Proof of Lemma 2. "To be disjoint with A" means "to be a subset of the complement of A"; so it remains to use the previous lemma and the fact that the complement of a set from the class Σ_n $[\Pi_n]$ belongs to the class Π_n $[\Sigma_n]$. The proof is complete. $\qquad\square$

Now we are ready to prove that all $\mathbf{0}^{(n-1)}$-enumerable sets belong to the class Σ_n. This is also proved by induction over n.

Let us begin with the first nontrivial case: Why does any $\mathbf{0}'$-enumerable set belong to Σ_2? (This can be explained using the criterion of $\mathbf{0}'$-computability given above, but it will be more instructive to use the general argument valid for all values of n.)

Suppose that a set A is $\mathbf{0}'$-enumerable. Then it is enumerable relative to a certain enumerable set B, i.e., it is enumerable relative to the characteristic function b of B. By the criterion we have proved above (Theorem 45, p. 74), this means that there exists an enumerable set Q of pairs of the form $\langle x, t \rangle$, where x is a number and t is a pattern, such that

$$x \in A \Leftrightarrow \exists t[(\langle x, t \rangle \in Q) \text{ and } (b \text{ extends } t)].$$

Without loss of generality we can assume that the patterns we consider here take only the values 0 and 1, because if t takes any other values, then it cannot be a part of the characteristic function of B and is of no significance to us. In terms of the set B, the condition "b extends t" reads as follows: B contains the set of arguments at which t takes the value 1 and is disjoint with the set of arguments at which t takes the value 0. Therefore, instead of patterns, we can speak about pairs of finite sets; then instead of Q we must consider the enumerable set P of triples of the form $\langle x, u, v \rangle$ and write:

$$x \in A \Leftrightarrow \exists u \exists v[(\langle x, u, v \rangle \in P) \text{ and } (B \text{ contains } D_u)$$

$$\text{and } (D_v \text{ is disjoint with } B)].$$

Now, instead of "B contains D_u" we can write "$u \in \text{Subset}(B)$", and instead of "D_v is disjoint with B" we can write "$v \in \text{Disjoint}(B)$". It remains to notice that all three properties joined by the conjunction "and" in the right-hand side belong to the class Σ_2 and even to lower classes. Namely, the first two belong to the class Σ_1, because P and B are enumerable (for the second property we apply Lemma 1). As to the third property, it belongs to the class Π_1 by Lemma 2. Therefore, their conjunction belongs to the class Σ_2, and the projection (the quantifiers $\exists u \exists v$) leaves us within this class. This completes the argument for $n = 2$.

Further, suppose that some set A is $\mathbf{0}''$-enumerable. By the definition, this means that A is enumerable with respect to a certain $\mathbf{0}'$-enumerable B. As we already know, B lies in Σ_2. From here on the reasoning can be simply repeated with all subscripts increased by 1. All subsequent values of n are considered similarly.   □

The theorem we have proved immediately implies the following corollary:

Theorem 57. *The intersection $\Sigma_n \cap \Pi_n$ coincides with the class of $\mathbf{0}^{(n-1)}$-decidable sets.*

Proof. Indeed, the relativized Post's Theorem (Theorem 2, p. 7) states that a set is X-decidable if and only if both this set and its complement are X-enumerable (here X is an arbitrary oracle).   □

Theorem 58. *The class $\Sigma_n \cup \Pi_n$ is a proper subset of the class $\Sigma_{n+1} \cap \Pi_{n+1}$.*

Proof. Recall that $\mathbf{0}^{(n)}$ is the degree of a set X m-complete in the class of $\mathbf{0}^{(n-1)}$-enumerable sets. Being m-complete in this class, X is not $\mathbf{0}^{(n-1)}$-decidable, that is, its complement is not $\mathbf{0}^{(n-1)}$-enumerable.

Therefore, by the previous theorem, X belongs to the class Σ_n, whereas its complement does not. On the other hand, the complement of X belongs to Π_n, but not to Σ_n. Now let us consider the combination of the set X with its complement, i.e., the set

$$Y = \{2n \mid n \in X\} \cup \{2n+1 \mid n \notin X\}.$$

Both X and the complement of X are m-reducible to Y, hence Y belongs neither to Σ_n nor to Π_n. At the same time, Y is obviously decidable with respect to X; therefore, by Theorem 57, Y belongs both to Σ_{n+1} and Π_{n+1}. □

4. Classification of sets in the hierarchy

It is interesting to find the position of a given specific set in the hierarchy described above. For instance, what can we say about the set of numbers of a given computable function in a Gödel numbering?

We have already said that the set of all numbers of all functions with nonempty domains is enumerable, i.e., belongs to the class Σ_1. Consequently, its complement, the set Z of all numbers of the empty function, belongs to the class Π_1. (The set Z cannot belong to the class Σ_1, because Z is undecidable; see Theorem 21, p. 29.)

Problem 69. Prove that the set of numbers of the empty function in any Gödel numbering is m-complete in the class Π_1.

What can we say about the numbers of other functions? For instance, what can we say about the set of numbers of the zero function (zero$(x) = 0$ for all x)? The following theorem gives a comprehensive answer to this question.

Theorem 59. (a) *Let U be a computable universal function for the class of computable functions. Then the set of all n such that $U_n = $ zero belongs to the class Π_2.* (b) *If, in addition, U is a Gödel universal function, then this set is m-complete in the class Π_2.*

It is essential that the universal function in statement (b) is a Gödel one: as we have seen on p. 30, there exists a computable numbering such that any computable function has only one number.

Proof. The property $U_n = $ zero can be rewritten as follows: for any k there exists a t such that *the computation of the value $U(n, k)$ terminates in t steps and returns* 0. The highlighted property is decidable, and is preceded by two quantifiers just of the desired type, which completes the proof of item (a).

Let us prove statement (b). Let P be an arbitrary set of the class Π_2. Then

$$x \in P \Leftrightarrow \forall y \exists z R(x, y, z)$$

for some decidable property R. Now we consider the function $S(x, y)$ computed by the following algorithm: searching through all natural numbers, we look for a number z such that $R(x, y, z)$ holds; once (and if) such a number is found, we send 0 to the output. Clearly, $S_x = $ zero if and only if $x \in P$. Using the assumption that U is a Gödel function, we can find a function s such that $U_{s(x)} = S_x$. This function reduces P to the set of all numbers of the zero function. □

What about other functions? For any computable universal function U and any computable function f, the set of all U-numbers of the function f is a Π_2-set. Even a stronger fact is valid: the property $U_m = U_n$ (integers m and n are numbers of the same function) is a Π_2-property of the pair $\langle m, n \rangle$, and, therefore, any of its sections (i.e., the set of numbers of some particular function) is all the more a Π_2-set. Indeed, the property $U_m = U_n$ can be formulated as follows: "for any x and t_1 there exists a t_2 such that *if the computation of $U(m, x)$ terminates in t_1 steps, then the computation of $U(n, x)$ terminates in t_2 steps with the same result, and vice versa*". The highlighted property is decidable, and is preceded by a Π_2-prefix.

It is possible to find out which functions have a Π_2-complete set of numbers: these are the functions with an infinite domain. If the domain of a function is finite, then the set of all its numbers is $\mathbf{0}'$-decidable (and so not Π_2-complete). Indeed, using an oracle for the halting problem, we can verify that the function is actually defined where it must be defined (and make sure that the values are correct). Then we can verify that it is undefined at all the remaining points (the search for a point that lies outside a given finite set and belongs to the function domain is an enumerable process, so we can check whether it succeeds by asking the $\mathbf{0}'$-oracle).

If the domain of a function f is infinite, then there exists an infinite decidable subset F of f's domain (Problem 12, p. 9). Now we can use essentially the same construction as for the zero function, but only inside F.

Problem 70. Complete this argument.

Problem 71. Show that the set of all numbers of all total functions (in a Gödel numbering) is Π_2-complete.

Problem 72. What is the lowest class of arithmetical hierarchy that contains the set of numbers of all functions with infinite domain? Is this set m-complete in this class?

Problem 73. Show that for any (not necessarily Gödel!) numbering the set T of all numbers of total functions is not enumerable. Furthermore, T has no enumerable subset that includes at least one number of each computable total function. (*Hint*: use the diagonal construction.)

H. Rogers ([11], section 14.8) gives results of this kind for many other properties of computable functions and enumerable sets. For instance, for any m-complete set K, the set of all its numbers is Π_2-complete. (From here on, speaking about numbers we mean a Gödel numbering of enumerable sets.) The set of numbers of all finite sets is Σ_2-complete. The set of numbers of the sets containing at least one number of an infinite set is Σ_3-complete. The set of numbers of all decidable sets is Σ_3-complete. The set of numbers of all the sets with finite complements is Σ_3-complete.

Problem 74. Prove these statements (or read the proofs in Rogers [11]).

Problem 75. Consider the quantifier $\exists^\infty x$ that means "there exist infinitely many x such that...". Show that all properties of the class Π_2 (and only they) are representable in the form

$$\exists^\infty x \,(\text{a decidable property}),$$

and that all properties of the class Σ_3 (and only they) are representable in the form

$$\exists^\infty x \,\forall y \,(\text{a decidable property}).$$

(A similar statement holds for higher classes as well, see Theorem XVIII in [**11**], section 14.8.)

Chapter 9

Turing Machines

1. Simple computational models: What do we need them for?

Thus far, when speaking about computations, we often referred to our experience with algorithms, programs, compilers, tracing, etc. This permitted us to skip details of algorithms on the pretext that the reader could easily restore them (or at least would take them for granted: after all, writing Pascal compilers in Pascal is not a popular hobby).

But in certain cases the commonsense approach does not suffice. For instance, imagine that we want to prove that some problem is algorithmically unsolvable, whereas the problem itself has nothing to do with algorithms. In this chapter we consider one such problem, the Thue, or word, problem in semigroups. The undecidability of such a problem is usually proved by reducing the Halting Problem to it. To this end, we simulate an arbitrary algorithm in terms of our problem (the example below will explain what we mean by that). In so doing, we must use some precise definition of an algorithm, and it is important to have as simple a definition (computational model) as possible to simplify the simulation.

Let us summarize our plan. We start with a class of computing devices with a relatively simple definition. They are known as Turing

machines and were introduced by Alan Turing. Then we will declare
that any computable function can be computed by a Turing machine.
Finally, we will show that the Halting Problem for Turing machines
can be reduced to the word problem in semigroups.

Another reason why simple computational models (various kinds
of Turing machines, random access machines, and many others) are
important has to do with computation complexity theory, which fo-
cuses on time and space bounds for a computation. But this subject
is beyond the scope of our book.

2. Turing machines: The definition

A *Turing machine* has a *tape* infinite in both directions; the tape is
divided into squares, or *cells*. Each cell holds a character from a fixed
finite set, called the *alphabet* of the machine. One character is singled
out and is called the "blank symbol" or "space". Initially the tape
is assumed to be completely blank (i.e., filled with blank symbols)
except for a finite part with the input string.

To change characters on the tape, a Turing machine uses a
read/write *head* that can move along the tape. At each step, the
head is positioned over one of the cells. Depending on the character
read by the head and the internal state, the machine decides what
to do, i.e., what character to write in the current cell and where to
move next (to the left, to the right, or just to stay in place). The
state of the machine is also changeable; we assume that the machine
has finitely many states or, in other words, a finite internal memory.
(The tape can be viewed as external memory; it is potentially infinite.) Finally, we must specify how the machine starts and when the
computation terminates.

The formal definition of a Turing machine includes the following
items:

- a finite set A (the *alphabet* or *tape alphabet*); its elements
  are called *characters*;

- a designated character $a_0 \in A$, called the *blank symbol* or
  space;

- a finite set S, whose elements are called *states*;

- a designated state $s_0 \in S$, called the *initial state*;

- a *transition table*, which determines the behavior of the machine as a function of its current state and current character (see below);

- a subset $F \subset S$, whose elements are called *terminal states*: when the machine attains any of these states, it stops.

The transition table assigns to each pair ⟨current state, current character⟩ a triple ⟨new state, new character, shift⟩. The "shift" here is one of the numbers -1 (left move), 0 (no move), or 1 (right move). Thus the transition table is a function of the type $S \times A \to S \times A \times \{-1, 0, 1\}$ defined on the pairs in which the current state is not terminal.

So how does this engine work? At each step, the situation is described by the machine's *configuration* which includes the tape contents (formally, the tape contents is an arbitrary mapping $\mathbb{Z} \to A$), the current position of the head (some integer), and the current state of the machine (an element of S). A configuration is transformed into the next one by natural rules: we look up the current state and tape character in the table, read the corresponding entry, and change the state of the machine and the character on the tape to new ones; then we move the head to the left, to the right, or leave it in place (by adding the "shift" to the "head position"). If the new state turns out to be terminal, the process terminates. Otherwise, everything is repeated again.

It remains to specify how the machine processes its input and what is its output. We will assume that the tape alphabet, in addition to the blank symbol, includes the characters 0 and 1 (and, possibly, some other characters). The input and output of the machine are finite zero-one sequences (bit strings). Initially, the input string is written on the blank tape, and the head is placed on its first character. Then the machine is put into the initial state and launched. If the computation terminates, then the output is the 0-1-string starting from the head position and delimited by any character distinct from 0 and 1.

Thus any Turing machine defines a (partial) function on binary strings. Each of these functions is said to be *computable by a Turing machine* or *Turing-computable*.

3. Turing machines: Discussion

Our definition involves many insignificant details that could be changed. For instance, the tape could be bounded on one side. Or we could supply the machine with two tapes (and two heads). We could allow the machine to either write a new character or move, but not both. We could confine the alphabet to, say, exactly 10 characters. We could require that in the end the tape should contain nothing except the output (all the remaining cells should be blank). All these modifications, as well as many others, do not change the class of Turing-computable functions.

However, some modifications can be quite dangerous. For instance, if we forbid the machine to move left, the situation will change dramatically: the tape will become practically useless, since we will no longer be able to return to our old records.

How can one distinguish the difference between harmless and dangerous modifications? Apparently, some experience in practical programming for Turing machines will be helpful. To have some practice, let us describe a machine that doubles the input string (makes a string XX from the input string X).

If the machine sees the blank symbol at the first step (the input string is blank), it immediately halts. If not, it memorizes the current character and marks it, e.g., with a bar (in addition to the characters 0 and 1, the alphabet contains their "barred versions" $\bar{0}$ and $\bar{1}$). Then the machine moves to the right until it reaches the first blank cell and writes the copy of the memorized character in it. Then it goes back to the mark. Once it arrives at the mark, it makes one step right, memorizes the next character, and so on, until the entire string is copied.

An experienced Turing machine programmer will readily discern fragments of the transition table behind this informal description. For instance, the words "it memorizes the character and moves right" imply that states are divided into two groups: one for the situations in which the character 0 is memorized, the other for 1, and the program that pushes the head to the first blank cell on the right is encoded inside each group.

But a more experienced programmer will see a bug in this description: we did not explain how the machine is supposed to learn that the whole string has been copied and it is time to stop: it cannot distinguish the copied characters from the original string. It is also clear how to patch the program: we must write some other special characters as the copies, say, $\tilde{0}$ and $\tilde{1}$; after the string is doubled, these characters must be replaced by 0 and 1.

Problem 76. Show that the inversion function i that rewrites a given string backwards (e.g., $i(001) = 100$) is Turing-computable.

Here is another example of informal reasoning: let us explain why additional characters distinct from 0, 1, and the blank symbol (space) are, in fact, unnecessary (i.e., the class of Turing-computable functions does not change if we restrict ourselves to these three characters). Consider a Turing machine M with a large alphabet of N characters. Let us construct a new machine that will simulate M using only the three "main" characters. Each cell on the tape of M corresponds to a k-cell block on the tape of the new machine. The block length k is chosen so as to be able to encode (by a string of k zeros, ones, and spaces) any of the N initial characters. In particular, the characters 0, 1, and space of the original alphabet will be encoded as 0 followed by $(k-1)$ spaces, 1 followed by $(k-1)$ spaces, and a group of k spaces, respectively.

To begin with, the neighboring characters of the input string must be moved apart at a distance of k, which can be done without additional characters. (The head goes to the rightmost input bit and carries it to the right, k cells away; then it comes back to the next bit and carries it together with the first one, and so on. The end of the string can be identified as the bit followed by at least k spaces.) Clearly, the procedure can be implemented using only a finite memory (depending on k).

Once the spaces are inserted, we are ready to simulate any computation performed by the machine M step by step, and this also requires only a finite memory (i.e., finitely many states), because we must take into account only a finite neighborhood of the head of the machine. In the end, the output string must be compressed back by deleting spaces between bits.

What functions can be computed by Turing machines? According to the *Turing Thesis*, any computable function is Turing-computable. Naturally, the meaning of this statement depends on what is understood by the term "computable function". If it is understood in a vague intuitive sense (as "a function that can be evaluated algorithmically", that is, "by clear unambiguous rules" or something like that), then, of course, a rigorous proof of the Turing Thesis is out of the question. The only thing we can say is that centuries-old practices from Euclid to Knuth have never encountered an example of an algorithm that could not be translated into a Turing machine program, and the like. However, below we give another argument (not too convincing, though).

But if we think of the term "computable" in the Turing Thesis as "computable by means of a Pascal program" and imagine for a while that the syntax and semantics of Pascal programs are defined well, then the Turing Thesis becomes a clear-cut statement that can be proved or disproved. Of course, such a proof must be based on a formal description of the syntax and semantics of Pascal, and so it was never carried out. However, for simpler computational models, proofs of this sort have actually been given. They are akin to proofs of correctness of lengthy programs; few are the volunteers to write such proofs, but even fewer are the volunteers to read them.

In conclusion, we present the above-mentioned informal argument in favor of Turing machine computability of any computable function. Suppose that you (or any other human being) can evaluate a certain function f for any given argument. Let us describe a Turing machine that simulates your work.

You will naturally use paper and pencil (with eraser), because the amount of information you can keep in your head is limited. Let us assume that you write on paper sheets of the same size: you have two stacks of sheets, one on each side of the current sheet; having completed your work with the current sheet, you can put it onto either stack and take the next working sheet from the top of the other stack.

To be able to discern the letters on the sheet, you will write them not too small, and so there are finitely many discernible states of the sheet. Therefore, we can assume that a sheet always holds only one character from an enormous, but finite, alphabet. Human memory is also finite and can be represented by a finite set of states. Furthermore, all your possible actions in any of your states and for any character on the current sheet can be entered into a table saying what you will write on the sheet after you process it, what your next state will be, and how you will rearrange the sheets. This table can be viewed as the transition table of our "human Turing machine" with a very large (but finite) alphabet and a large (but finite) number of internal states.

4. The word problem

We are going to use Turing machines to prove the unsolvability of a certain algorithmic problem concerning words and their transformations.

Recall that an *alphabet* is a finite set whose elements are called *characters*; finite sequences of characters are called *strings* (or words).

Let us fix an alphabet A. An arbitrary expression of the form $P \to Q$, where P and Q are strings over the given alphabet (we assume that the arrow symbol itself does not belong to A) is called a (*string-rewriting* or *transformation*) *rule* (another term you may encounter is "production"). Any (unordered) finite set of transformation rules is called a *string-rewriting system* or *semi-Thue system*. So how do we use these rules in string rewriting? We say that a rule $P \to Q$ is *applicable* to a string X if P is a *substring* of X, i.e., if $X = RPS$ for some strings R and S. In this case we are allowed to replace P by Q and get RQS. There can be several occurrences of P in X; then we can apply the same rule to the same string in a number of ways. The system can contain several different rules applicable to a given string; then we can apply any of them. After that, the same or another rule of the system can be applied to the string again, and so on.

Let us repeat this definition more formally. We say that a string X can be *transformed* into a string Y by the rules of a system I if there exists a finite sequence of strings

$$X = Z_0, Z_1, Z_2, \ldots, Z_{k-1}, Z_k = Y$$

in which every string Z_i is obtained from the preceding string Z_{i-1} by one of the rules from I: there is a rule $P \to Q$ in I such that $Z_{i-1} = RPS$ and $Z_i = RQS$ for some strings R and S.

Thus, each set of rules I determines a certain set of pairs of strings, namely, the set of pairs $\langle X, Y \rangle$ such that X can be transformed into Y by the rules from I.

Theorem 60. *For any system I the corresponding set of pairs $P(I)$ is enumerable. There exists a system I such that the set $P(I)$ is undecidable.*

We prove this theorem in the following section. The first statement is easy: the set of all the chains $Z_0 \to Z_1 \to \cdots \to Z_k$ complying with the rules of the system is decidable and hence enumerable. Taking only the first and last strings from these chains, we obtain an enumeration of the set $P(I)$.

It remains to construct an example of the undecidable string-rewriting system I (i.e., a system with an undecidable set $P(I)$). To this end, we will show that any Turing machine can be simulated in terms of string-rewriting systems and then take the system corresponding to a machine with undecidable Halting Problem.

5. Simulation of Turing machines

Theorem 61. *Let M be a Turing machine whose tape alphabet includes the characters 0 and 1. Then we can construct a string-rewriting system I (whose alphabet includes 0, 1, the brackets [,] and, possibly, some other characters) with the following property: for any two binary strings X and Y, the machine M run on the input X yields the output Y if and only if the string $[X]$ can be transformed into the string Y by the rules of the system I.*

Recall that the output of the machine is defined as the maximal string of zeros and ones read to the right of the final head position.

Notice also that the alphabet of the system I is assumed to contain, together with the characters 0 and 1, additional characters [and] and, possibly, other characters.

Problem 77. Show that the auxiliary characters [and] are indispensable: the theorem will not hold if we replace the string $[X]$ in it by X. (*Hint*: if a string Y can be obtained from X by the rules of the system, then the string PYQ can be obtained from the string PXQ by the same rules.)

Proof. The idea of the simulation is as follows. We encode all configurations of the Turing machine (configurations include the tape contents, the head position, and the state) by strings in such a way that every single operation of the machine corresponds to the application of a certain rule from the system I.

Let us explain the encoding in more detail. The configuration

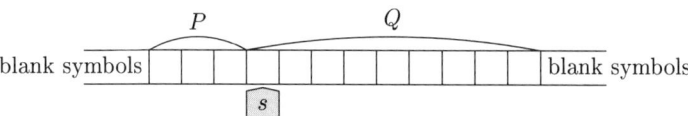

is encoded by the string $[PsQ]$. Thus the alphabet A of our system will consist of all tape characters of the Turing machine including the blank symbol (denoted by "_"), all states of the machine (we assume that the set of states is disjoint with the tape alphabet), and the special characters [and]. Notice that the encoding is not unique because the string P can have leading blanks and Q can have trailing blanks. For instance, if a, b, and c are characters of the alphabet, and s is a state, then the string [absc] encodes the configuration with the machine's state s, the tape contents ..._abc_..., and the head positioned against c. The same configuration can be encoded, e.g., by the strings [_absc] and [absc_]. A few other examples of configuration encoding: the string [sabc] encodes the state s, the tape ..._abc_..., and the head against a; the string [abcs] corresponds to the tape contents ..._abc_... with the head to the right of c; and the string [s] corresponds to the blank tape.

Now we must write the rules of the system simulating M. Our system will have the following property: for each configuration C, encoded by a string E, there exists a unique rule that can be applied to E; this rule produces the string E' that encodes the next configuration C' of M. Then each transformation by a single rule will correspond to a single step of the machine computation.

In fact, what we are going to do is translate the transition table of the Turing machine into the language of rules. For example, suppose that the table contains the instruction "if the current character is x and the current state is s, switch to the state s', print the character x', and stay in place" (in other words, the transition function takes $\langle s, x \rangle$ to $\langle s', x', 0 \rangle$). Then we add to our system the rule

$$sx \to s'x'.$$

The instruction "if the current character and state are x and s, switch to the state s', print the character x', and move left" (the transition from $\langle s, x \rangle$ to $\langle s', x', -1 \rangle$) generates the rules

$$\alpha sx \to s'\alpha x'$$

for all characters α of the tape alphabet.

The transition from $\langle s, x \rangle$ to $\langle s', x', 1 \rangle$ (right move) generates the rule

$$sx \to x's'.$$

But we also must take care of the situations in which the string P, or Q, or both (see the figure) are blank. Here we need these rules:

the transition	generates the rule
from $\langle s, x \rangle$ to $\langle s', x', -1 \rangle$	$[sx \to [s'\_x'$
from $\langle s, \_ \rangle$ to $\langle s', x', 0 \rangle$	$s] \to s'x']$
from $\langle s, \_ \rangle$ to $\langle s', x', -1 \rangle$	$\alpha s] \to s'\alpha x']$
	$[s] \to [s'\_x']$
from $\langle s, \_ \rangle$ to $\langle s', r', 1 \rangle$	$s] \to x's']$

The special cases described in this table are the left move for the blank string P and the blank symbol under the head with the blank string Q.

Applying these rules one by one, we simulate the computation by the Turing machine. Yet it remains to "preprocess the input" and to "postprocess the output". We will start with the second, more complicated operation. It is intended to extract the result of the computation from the code of the terminal configuration of M. If the terminal state of M is s, then this code is of the form $[PsQ]$. That is, we must delete the string P and the opening bracket from the code, single out the maximal 0-1-prefix of the string Q, and erase the rest of this string. This can be done as follows.

We introduce an additional character $\triangleleft$, and the rules

$$s \to \triangleleft \text{ (for each terminal state } s),$$
$$\alpha \triangleleft \to \triangleleft \text{ (for each character } \alpha \neq [),$$
$$[\triangleleft \to \triangleright.$$

By these rules, the character $\triangleleft$ will replace the terminal state symbol s, then it will gobble all characters to the left of itself up to the bracket, then swallow the bracket and at the same time flip into the new character $\triangleright$. This character obeys the following rules:

$$\triangleright 0 \to 0 \triangleright,$$
$$\triangleright 1 \to 1 \triangleright,$$
$$\triangleright \alpha \to \triangledown \alpha$$

(the last rule is applicable to all characters α except 0 and 1). The character $\triangleright$ walks through the result of computation (the 0-1-string to the right of itself) and turns into the character $\triangledown$ at the string's end. Finally, the third triangular character wipes out everything to the right of itself and then annihilates together with the closing bracket. The rules are these:

$$\triangledown \alpha \to \triangledown \text{ (for all characters } \alpha \neq]),$$
$$\triangledown] \to \Lambda \text{ (Λ denotes the blank string).}$$

These rules enable us to extract the output of the machine from the code of the terminal configuration. Now it is safe to say that the machine yields the output Y on the input X if and only if the string Y can be obtained by the rules above from the string $[s_0 X]$. The only distinction from the statement of the theorem is in the new character s_0. This is, however, easily corrected: we add the new character $['$, the rule $[\to ['s_0$, and replace $[$ in all the other rules by $['$. (This is just the above-mentioned "preprocessing" of the input.)

Now, at last, everything has been brought into exact correspondence with the statement of the theorem, and we can declare the proof completed. (Alas, an accurate implementation of an almost obvious idea often requires explanations of numerous details.) □

Now we are ready to construct the undecidable string-rewriting system (Theorem 60), i.e., a string-rewriting system I with an undecidable set $P(I)$.

Let us take an enumerable undecidable set K. Consider the Turing machine that halts and returns the blank string Λ on any input from K, and does not halt on any input not in K. (Since K is an enumerable set, the function defined on K that takes the value Λ on any input in K is computable, and so, by the Turing Thesis, it is Turing-computable.)

Let us construct the string-rewriting system that simulates this machine as described above. For this system, there is no algorithm that could find out whether an arbitrary given string U can be transformed into another given string V. Indeed, any such algorithm (applied to the strings $[X]$ and Λ) would tell us whether the string X belongs to the undecidable set K or not.

6. Thue systems

It turns out that the argument of the previous section can be slightly strengthened. If each rule $X \rightarrow Y$ in a given set of rules is accompanied by the inverse rule $Y \rightarrow X$, then this set is called a *Thue system* (hence the term semi-Thue systems for the sets of rules considered above).

Theorem 62. *There exists a Thue system such that no algorithm can determine whether one string can be obtained from another by the rules of this system.*

Proof. The proof uses the above construction of the undecidable string-rewriting system I (that simulates a Turing machine M with undecidable Halting Problem) with minor modifications. First, the rule $\nabla] \rightarrow \Lambda$ is replaced by the rule $\nabla] \rightarrow \star$ (soon we will see what this is done for). Second, for each rule from I, we add the inverse one.

The new system I' thus obtained simulates the Turing machine M as well:

Lemma. *A binary string Y is the output of the machine on the input binary string X if and only if the string $[X]$ can be transformed into the string $Y\star$ by the rules of the system I'.*

Proof of the Lemma. If the inverse rules were not added, then I' would be no different from the system I constructed in the previous section (except for the last step, where the star remains, but this is obviously inessential). Therefore, it will suffice to show that if $[X]$ can be transformed into the string $Y\star$ by the I'-rules, then this transformation can be performed without applying the inverse rules, only by the direct ones.

This is proved as follows. Let us call the following characters "active": the $[$ character (which, as we know, is replaced at the very first step), all states of the machine, $\triangleright$, $\triangleleft$, $\triangledown$, and $\star$. Then each rule of our system contains exactly one active character on either side of the arrow. Hence all the strings in any sequence of direct and inverse rules transforming $[X]$ into $Y\star$ contain one active character each.

Now let us notice that there is at most one direct rule applicable to a string with a single active character. (This can be verified by inspecting all the rules; the reason is that the rules simulate a deterministic Turing machine in which each configuration is uniquely defined by the previous one.) Therefore, we can delete all inverse rules from the sequence that transforms $[X]$ into $Y\star$.

Indeed, consider the last occurrence of an inverse rule in this sequence of transformations. It cannot be the last in the sequence, because inverse rules do not generate the character $\star$. Therefore, it is followed by the application of a certain direct rule. But we already have one applicable direct rule: this is the rule whose inversion is the inverse rule in question. By uniqueness, we see that our inverse rule is followed by its direct counterpart. Therefore, these rules cancel out to yield a shorter sequence in which we can find the last inverse rule again, etc. This completes the proof of the Lemma. $\qquad\square$

The Lemma immediately implies the existence of undecidable Thue systems, which proves the statement of the main theorem of this section.   □

7. Semigroups, generators, and relations

The property of Thue systems proved in the previous section (Theorem 62) can be translated into algebraic language. (We are not going to prove any essentially new statements; this is only a translation.) Recall some relevant algebraic material.

A *semigroup* is an arbitrary nonempty set G with an associative operation, which will be written as multiplication. All the semigroups we consider also have a unit element 1 (such that $1 \times x = x \times 1 = x$ for all $x \in G$). Such semigroups are usually called monoids, but we do not use this term. If A is a subset of a semigroup G and any element of G can be represented as the product of elements of A, then we say that A *generates* G; elements of A are called *generators*. (The empty product is also allowed; it is assumed to be equal to 1.) A semigroup is said to be *finitely generated* if it has a finite set of generators.

Let $A = \{a_1, \ldots, a_n\}$ be an alphabet. Then the set of all strings over the alphabet A with string *concatenation* (i.e., juxtaposition of two strings) as multiplication is a semigroup. The unit element here is the blank string. Obviously, the characters $a_1, \ldots, a_n$ are generators of this semigroup. (It is more correct to speak about one-character strings as generators.) This semigroup is called the *free semigroup* with generators $a_1, \ldots, a_n$ and is denoted by $\mathcal{F}(a_1, \ldots, a_n)$.

Let G be an arbitrary semigroup, and let $g_1, \ldots, g_n$ be its arbitrary elements. Then there exists a unique homomorphism h of the semigroup $\mathcal{F}(a_1, \ldots, a_n)$ to the semigroup G such that $h(a_i) = g_i$. (A *homomorphism* of semigroups is a mapping that takes the product of elements into the product of their images and the unit element of one semigroup into that of the other.) It transforms any string of the characters a_i into the product of the corresponding elements g_i; the image of the blank string is the unit of G. Obviously, the image of $\mathcal{F}(a_1, \ldots, a_n)$ under this homomorphism coincides with the entire semigroup G if and only if the elements $g_1, \ldots, g_n$ are generators of G.

Equations of the form $X = Y$, where X and Y are elements of the free semigroup $\mathcal{F}(a_1, \ldots, a_n)$, that is, strings over the alphabet A, are called *relations*. We say that a relation $X = Y$ holds in a semigroup G with marked elements $g_1, \ldots, g_n$ if the images of the strings X and Y under the homomorphism described above, that is, the products of the corresponding elements g_i, are equal. Suppose that we have a set of relations $X_1 = Y_1, \ldots, X_k = Y_k$. We will consider various semigroups G with n marked elements in which all these relations hold and the marked elements are generators. (For instance, one of these semigroups is the semigroup consisting of a single element, the unit.) As we will see soon, among these semigroups there is a "maximal" one, with the smallest number of relations.

Let us introduce an equivalence relation on A-strings by setting $P \equiv Q$ if P can be transformed into Q in the Thue system with the rules $X_1 \leftrightarrow Y_1, \ldots, X_k \leftrightarrow Y_k$. (In other words, we are allowed to replace the substring X_i of any string by the substring Y_i and vice versa.) Obviously, this relation is indeed an equivalence. Notice that if $P \equiv Q$, then for an arbitrary string R, $PR \equiv QR$ and $RP \equiv RQ$ (we can add R on the right or on the left of each string in the sequence of transformations). Consider equivalence classes with respect to this relation. We can define multiplication of classes by setting the product of two classes containing strings P and Q to be the class containing their concatenation PQ. The above-mentioned property of equivalence guarantees that the product is well defined (the class of the product does not depend on the choice of representatives in the factors). Thus we obtain a semigroup G with the class of the blank string as the unit element and the equivalence classes $g_i - [a_i]$ of one-character strings as generators. This semigroup is denoted by

$$\mathcal{F}(a_1, \ldots, a_n)/(X_1 = Y_1, \ldots, X_k = Y_k)$$

and is called the *semigroup with generators* $a_1, \ldots, a_k$ *and relations* $X_1 = Y_1, \ldots, X_k = Y_k$. Clearly, the initial relations $X_i = Y_i$ hold in the semigroup G. Also, we can readily see that any relation in it is a corollary of the initial ones:

Theorem 63. *If the relation* $X = Y$ *holds in the semigroup*

$$\mathcal{F}(a_1, \ldots, a_n)/(X_1 = Y_1, \ldots, X_k = Y_k),$$

then it holds in any semigroup G with marked elements $g_1, \ldots, g_n$ in which all the relations $X_i = Y_i$ hold.

Proof. We have assigned to the strings X and Y their equivalence classes with respect to the relation specified above so that Y can be obtained from X by the rules of the Thue system. But none of these rules changes the value of a string in any semigroup satisfying the relations $X_i = Y_i$; so in any such semigroup the elements corresponding to X and Y will be the same. $\qquad\square$

Problem 78. What is the semigroup with two generators a_1 and a_2 and the relations $a_1 a_2 = \Lambda$, $a_2 a_1 = \Lambda$?

Problem 79. What is the semigroup with two generators a_1 and a_2 and the relation $a_1 a_2 = a_2 a_1$?

Problem 80. What is the semigroup with two generators a_1 and a_2 and the relations $a_1 a_1 = \Lambda$, $a_2 a_2 = \Lambda$, $a_1 a_2 = a_2 a_1$?

Problem 81. What is the semigroup with two generators a_1 and a_2 and the relations $a_1 a_1 = \Lambda$, $a_2 a_2 a_2 = \Lambda$, $a_1 a_2 = a_2 a_2 a_1$?

Now we are ready to reformulate the statement about undecidable Thue systems in terms of semigroups.

Theorem 64. *There exists a semigroup with finitely many generators and finitely many relations in which the problem of verifying the equality of two strings of generators is algorithmically unsolvable (there is no algorithm to determine whether or not two given strings are equal in this semigroup).*

Proof. By the definition, the equality of two strings of generators means that one of them can be transformed into the other by the rules of the Thue system; so this theorem is a reformulation of Theorem 62 (p. 118). $\qquad\square$

This theorem was proved around 1947 independently by Emil Post and Andrei Markov (Jr.); soon after that Pyotr Novikov and later William Boone strengthened it: they have constructed an example of a group (rather than a semigroup!) with finitely many generators and relations for which the problem of equality of two strings of generators is undecidable.

Chapter 10

Arithmeticity of Computable Functions

1. Programs with a finite number of variables

We are going to show that the graph of any computable function is an arithmetical set, that is, can be specified by an arithmetical formula. It will be convenient to do this using a model other than Turing machines. This model can conventionally be called "machines with finitely many registers".

A program for such a machine uses a finite number of variables whose values are natural numbers. (These numbers can be arbitrarily large, so, actually, the machine memory is unbounded.) A program consists of numbered commands of one of the following forms each (where a and b are some variables):

- a:=0
- a:=b
- a:=b+1
- a:=b-1
- goto ⟨number⟩
- if a=0 then goto ⟨number1⟩ else goto ⟨number2⟩
- stop

For the reader unfamiliar with the `goto` command, we explain how the `if` command is executed: if the value of the variable a is equal to zero, then the command to be executed next is the one with the number specified after `then`; if a is not equal to zero, then we proceed to the command with the number specified after `else`. The unconditional command `goto` always transfers control to the command with the number it specifies.

Since we assume that the values of variables are natural numbers (nonnegative integers), we set the difference $0 - 1$ equal to 0 (however, this is not that important; such an event could be regarded as error condition).

By the `stop` command, the program halts.

As in the case of Turing machines, it will be useful to gain some programming experience. Let us program the addition of two numbers. The program must put the sum of the numbers a and b in the variable c. The corresponding Pascal program would look like this:

```
c:=a;
{invariant: answer=sum of the current values of c and b}
while b<>0 do begin
    c:=c+1;
    b:=b-1;
end;
```

The program for our machine is obtained by simulating the loop using the `goto` statements:

```
1 c:=a
2 if b=0 then goto 6 else goto 3
3 c:=c+1
4 b:=b-1
5 goto 2
6 stop
```

Now it is clear how to write programs for subtraction, multiplication (iterated addition), division, remainder (`mod` operation), exponentiation, testing for primality, computing the nth prime number, etc. In general, this language is more customary as compared to Turing machines, and for this reason it is easier to believe that all algorithms can be programmed in it.

The only thing it really lacks is arrays. But we can easily override this deficiency with available arbitrarily large numbers, because (as is common in the general theory of algorithms) we do not care about the number of operations. Instead of a bit array, we can store the number whose binary notation coincides with this array. Arrays of numbers can be encoded by one number. For example, we can store a sequence $\langle a, b, c, d, e \rangle$ as the number $2^a 3^b 5^c 7^d 11^e$. Then the commands a[i]:=b and b:=a[i] will be replaced by small programs involving, among others, the variables a, b, i. (In particular, these programs will include the computation of the nth prime for a given n.)

It is easy to give the definition of computable functions in this model. A program that evaluates a function f will have two variables x and y (and, possibly, some others). To find $f(n)$ for a given number n, we put n in the variable x and initialize to zero all the other variables. Then we run the program. If it does not halt, then f is undefined at n. If the program halts, then the value $f(n)$ is found in the variable y. A function is said to be computable (in this model) if there exists a program that computes it.

As usual, many details in this definition are inessential. Some commands could have been added (for instance, addition), others removed (say, using a little trick we can do without copying).

Problem 82. Show that the class of computable functions will remain the same if we remove the copy command (a:=b) from the definition.

The following fact is even more surprising. (It will not seem that strange, though, if we recall that a whole array can be squeezed into one variable.)

Problem 83. Prove that the number of variables can be limited by a fixed value, say, 100, without changing the class of computable functions.

2. Turing machines and programs

The computation model constructed above is not weaker than Turing machines in the following sense:

Theorem 65. *Any Turing-computable function can be computed by a program with finitely many variables.*

We should specify what we mean, because for a Turing machine the initial data and the result are binary strings, whereas for a program with finitely many variables they are natural numbers. We identify strings Λ (blank string), 0, 1, 00, 01,... with the numbers 0, 1, 2, 3, 4,... (to obtain a string from a number, we add 1 to the number, then convert the sum into the binary notation and trim the leading bit 1).

Proof. As before, we give only an approximate description of how to construct a program with finitely many variables that computes the same function as a given Turing machine. First of all we must encode the configuration of the Turing machine by integers. This can be done, for instance, by assigning four integers to each configuration: the number of the current state, the number of the current character (the one under the machine's head), the code of the tape contents to the left of the head, and the code of the tape contents to the right of the head.

To choose the most suitable way of encoding the two halves of the tape contents, we notice that the Turing machine treats them as stacks. (A stack is a data structure that resembles a pile of paper sheets. You can put a sheet on its top, get the top sheet, and check if there are other sheets.) Indeed, the right shift of the head can be thought of as moving the top element of the right-hand stack to the left-hand stack; the left shift works the other way around. Stacks are easily simulated by numbers: for instance, if the characters stored in a stack are 0 and 1, then pushing 0 or 1 is represented by the operation $x \mapsto 2x$ or $x \mapsto 2x + 1$, respectively, while removing the top element corresponds to the division by 2. In other words, we think of the binary notation of a number as a stack with the top element at its right end, in the lowest bit. In the same way, we can use the

k-ary notation to represent a stack with k possible characters in each position.

Now one can represent the basic loop of the Turing machine as a program that changes the four above-mentioned numbers (the current character, state, left stack, and right stack). However, a few things should be taken care of.

First, stacks are finite, whereas the tape is infinite; so let us agree that once the stack is emptied, we automatically put the blank symbol in it. So the infinite blank tails of the tape will reside in the stacks virtually.

Second, recall that we have agreed to identify binary strings (used as the inputs of the Turing machine) with their codes stored in the variables of our program. Therefore, after receiving the code of the input string, we must disassemble it into characters and put these characters one after another in the stack (the program and the machine use different number systems, so we cannot simply rewrite strings as they are). Similar problems arise when the output (a part of the right-hand stack contents) is converted into the corresponding number, but all of them can be easily resolved, and we will skip the details.   □

The converse statement is also true:

Theorem 66. *Any function computable by a program with finitely many variables is Turing machine computable.*

Proof. We must simulate the computation carried out by a program by means of a Turing machine. Suppose that the values of variables are written on the tape (in the binary notation) and are separated by a special character. Then the machine can find any variable moving from the beginning of the tape and counting separating characters, process this variable, and then come back to the beginning. (It is not necessary to keep the number of the executed command on the tape, since there are only finitely many commands, and the machine can store the current command number as part of its state.) The add-one and subtract-one operations are easily performed in the binary notation (when moving from right to left). We should only take into account that the number length may change after these operations.

Addition may increase the length; then we must clear an extra cell to the left of the number by shifting all the characters one position to the right of the head. (After subtraction, the characters will possibly have to be shifted to the left.) Clearly, this can also be easily performed by our Turing machine.

If numbers are written in binary notation, then the input–output recoding boils down to filling with zeros the other variables before the computation starts and landing the head on the terminal tape cell in the end.   □

3. Computable functions are arithmetical

Now we will prove that functions computable by programs with finitely many variables are arithmetical, that is, their graphs are arithmetical sets. In this section we assume again that the reader is familiar with basic logical notation. We will consider *arithmetical formulas* containing nonnegative-integer variables, the equality relation, the constants 0 and 1, operations of addition and multiplication, the logical connectives (AND, OR, NOT), and the quantifiers "for all" and "there exists". Formally, we will deal with the first-order language that contains one binary predicate symbol (equality), two constants (0 and 1), and two binary functional symbols (addition and multiplication). By saying that such a formula is true, we mean that it is true in the standard interpretation over the set $\mathbb{N}$ of natural numbers.

A set $A \subset \mathbb{N}^k$ is said to be *arithmetical* if there exists an arithmetical formula α with parameters $x_1, \ldots, x_k$ that represents it in the following sense: $\langle n_1, \ldots, n_k \rangle \in A$ if and only if the formula α is true for the parameter values $x_1 = n_1, \ldots, x_k = n_k$.

Theorem 67. *The graph of any function computable by a program with finitely many variables is an arithmetical set.*

Proof. Let $f \colon \mathbb{N} \to \mathbb{N}$ be the function computed by a program P with finitely many variables $k_1, \ldots, k_N$. We will assume that the input and output variables are k_1 and k_2, respectively. We want to write a formula with two variables x and y that would be true if and only if $y = f(x)$. A state of a program with finitely many variables is completely specified by the values of variables and the current command number (the corresponding register in processors is often called the *program counter*). It can be readily seen that the correspondence between two consecutive states of a program with finitely many variables is arithmetical. That is, one can write an arithmetical formula

$$\text{Step}(s_1, \ldots, s_N, p, s'_1, \ldots, s'_N, p')$$

with $2N + 2$ variables which says that one step of the program P takes it from the state with variables equal to $s_1, \ldots, s_N$ and the program counter equal to p into the state with variables equal to $s'_1, \ldots, s'_n$ and the program counter equal to p'. (We can and will assume that the value $p' = 0$ indicates the program halt.) This formula is obtained as the conjunction of individual statements corresponding to each line of the program. For instance, suppose that the 7th line of the program is of the form $k_2 := k_3$. Then the conjunction will include a term of the form

$$(p = 7) \Rightarrow ((s'_1 = s_1) \wedge (s'_2 = s_3) \wedge (s'_3 = s_3) \wedge \ldots$$
$$\ldots \wedge (s'_N = s_N) \wedge (p' = 8)).$$

A string with a conditional branch like

```
3   if k₅=0 then goto 17 else goto 33
```

will be represented in the formula by two conjunctive terms (for each of the two cases of transfer):

$$((p = 3) \wedge (s_5 = 0)) \Rightarrow ((s'_1 = s_1) \wedge \ldots \wedge (s'_N = s_N) \wedge (p' = 17))$$

and

$$((p = 3) \wedge (s_5 \neq 0)) \Rightarrow ((s'_1 = s_1) \wedge \ldots \wedge (s'_N = s_N) \wedge (p' = 33)).$$

Also, we must add the halting condition: for $p = 0$ all the variables preserve their values on the next step, and the program counter remains equal to zero ($p' = 0$).

Thus it is not difficult to prove that a single program step is arithmetical. But a computation is a *sequence* of steps that starts at the initial state, ends at a state satisfying the halting condition (with zero program counter), and such that each step in it is correct. So we still have to answer the main question: how do we write a formula that would express the existence of such a sequence? The difficulty lies in that we have to write something like a variable number of existential quantifiers, or the quantifier "there exists a finite sequence of natural numbers".

This is done by means of the technique traditionally called *Gödel's β-function*. This is what we mean:

Lemma 1. *For any k, there exists an arbitrarily large positive integer b such that the first k terms of the sequence $b+1, 2b+1, 3b+1, \ldots$ are pairwise coprime.*

Proof. Any common prime divisor p of two terms of the sequence is a divisor of their difference, i.e., the number lb for $0 < l < k$; taking b to be a multiple of $k!$, we ensure that p divides the number b, whereas all terms of our sequence are coprime with b. This completes the proof of Lemma 1.   □

Lemma 2. *For any sequence $x_0, x_1, \ldots, x_n$ of natural numbers, one can find numbers a and b such that x_i is the remainder of a when divided by $b(i + 1) + 1$.*

Proof. By the previous lemma, we can assume that the divisors $b(i+1)+1$ are coprime (and arbitrarily large). So it remains to apply the Chinese Remainder Theorem, which states that if positive integers $d_1, \ldots, d_k$ are coprime, then we can choose an integer u that yields any given set of remainders upon division by the numbers $d_1, \ldots, d_k$.

Indeed, there are $d_1 d_2 \cdots d_k$ such tuples (since the remainders upon division by d_i are the numbers from 0 to $d_i - 1$). All the numbers $u = 0, 1, \ldots, d_1 d_2 \cdots d_k - 1$ yield different sets of remainders (if two of these numbers, u' and u'', yield the same tuple of remainders, then their difference is divisible by all the numbers d_i, which is impossible because the numbers d_i are coprime, and the difference is less than their product). Now recall that there are equally many numbers u and tuples of remainders; hence any tuple can be represented as the remainders of one of the numbers u.

This completes the proof of Lemma 2.   □

Lemma 2 shows that a sequence of an arbitrary length can be encoded by three numbers a, b, and n. Thus, in a certain sense, we can replace the "formula"

$$\exists\langle x_0, \ldots, x_n\rangle (\forall i \leq n)[\ldots x_i \ldots]$$

(which actually is not a valid formula, since it contains the quantifier over finite sequences) by the formula

$$\exists a \, \exists b \, \exists n \, (\forall i \leq n)[\ldots (\text{the remainder of } a$$
$$\text{when divided by } b(i+1)+1)\ldots].$$

We will write the remainder of a upon division by $b(i+1)+1$ as $\beta(a, b, i)$ (hence the term "beta-function").

Returning to the program P with finitely many variables $k_1, \ldots, k_N$ and the function f it computes, we can write any statement of the form $f(x) = y$ as follows: there exist a number n of steps and numbers $a_1, b_1, a_2, b_2, \ldots, a_N, b_N, a, b$ such that

- $\beta(a_1, b_1, 0), \ldots, \beta(a_N, b_N, 0)$ are the correct initial values of variables (the first one being equal to x, all the rest to 0); $\beta(a, b, 0)$ is the correct initial value of the program counter, i.e., $\beta(a, b, 0) = 1$.

- for any i from 0 to $n-1$ we have

  $$\text{Step}(\beta(a_1, b_1, i), \ldots, \beta(a_N, b_N, i), \beta(a, b, i),$$
  $$\beta(a_1, b_1, i+1), \ldots, \beta(a_N, b_N, i+1), \beta(a, b, i+1)),$$

  that is, each transition complies with the program;

- $\beta(a_2, b_2, n) = y$ (the value of the output variable k_2 at the end of the computation is equal to y) and $\beta(a, b, n) = 0$ (the value of the program counter at the end of the computation is equal to 0, which corresponds to the program halt by our agreement).

This completes the proof of the arithmeticity of functions computable by machines with finitely many variables. $\qquad \square$

Recalling Theorem 65, we conclude that any Turing-computable function is arithmetical. Using the Turing Thesis, we conclude that the graph of any computable function is an arithmetical set.

4. Tarski and Gödel's Theorems

Thus, the graphs of computable functions are arithmetical. It follows immediately that decidable and enumerable sets are arithmetical as well. This justifies the name "arithmetical hierarchy" for the classes Σ_n and Π_n:

Theorem 68. *Any set that belongs to one of the classes Σ_n or Π_n (for any n) is arithmetical (i.e., the membership in this set can be expressed by an arithmetical formula).*

Proof. After we have proved that computable functions are arithmetical, everything is clear: sets of the classes Σ_n and Π_n are obtained by quantification of decidable predicates, and decidable predicates are arithmetical. □

The converse statement is also true:

Theorem 69. *Any arithmetical set belongs to the class Σ_n or Π_n for a certain n (and, of course, for all greater values).*

Proof. Let us convert the formula that defines an arithmetical set to the prenex normal form (by moving all the quantifiers to the left of the formula). Clearly, the part without quantifiers defines a decidable set; therefore, the initial set belongs to one of the classes Σ_n or Π_n.

Instead of using the prenex normal form, we could apply induction on the formula length and refer to the fact that the intersection, union, and complement, as well as projection, do not take us outside the arithmetical hierarchy (the union of all the classes Σ_n and Π_n). □

Now let us consider the set T consisting of all true arithmetical formulas without parameters (to be more exact, the set of all their numbers in a certain computable numbering of all formulas).

Theorem 70. *Any arithmetical set is m-reducible to the set T.*

Proof. This statement is almost obvious. Let A be an arbitrary arithmetical set. Let $\alpha(x)$ be a unary formula that expresses the membership in A. This means that $\alpha(n)$ is true if and only if n

belongs to A. Then the computable function $n \mapsto$ (the number of the formula obtained by substituting the constant n into $\alpha(x)$) m-reduces A to T. □

Theorem 71. *The set T is not arithmetical.*

Proof. The background we have built up makes this statement obvious: if the set T were arithmetical, then it would lie in one of the Σ_n classes. Then, since any arithmetical set reduces to T, by Theorem 53 all arithmetical sets would belong to this class. But we know that sets of higher classes of the hierarchy are also arithmetical, although they do not all belong to Σ_n. □

This fact is called *Tarski's Theorem*. It can be read as follows: "the set of arithmetical truths is not arithmetical". Or, "the notion of arithmetical truth cannot be defined in arithmetical terms".

Theorem 72. *The set T of arithmetical truths is not enumerable.*

Proof. Indeed, any enumerable set is arithmetical. □

This statement is called *Gödel's Incompleteness Theorem*. It can be reformulated as follows: if a formal system generating arithmetical formulas (i.e., an algorithm that enumerates a certain set of such formulas) is *sound* (i.e., it cannot generate a false formula), then it is *incomplete* (it cannot generate a certain true formula). (And any complete system is not sound.)

Problem 84. Show that for any N the set of all true closed arithmetical formulas with at most N quantifiers is arithmetical.

Problem 85. Formulate and prove a similar statement for formulas of bounded quantifier depth (the quantifier depth is the number of nested quantifiers; it is equal to the greatest length of a chain of quantifiers each of which belongs to the domain of its left neighbor), and for formulas with a bounded number of quantifier changes in the prenex normal form.

5. Direct proof of Tarski and Gödel's Theorems

We have obtained Tarski and Gödel's Theorems as simple conse-
quences of definitions and facts related to the theory of algorithms.
This allows us to understand better the place of these theorems in the
general context of mathematical logic and the theory of algorithms.
On the other hand, it would be instructive to convert these arguments
into more direct proofs. Such proofs are given below.

We start the proof of Tarski's Theorem by assuming that the set
T of numbers of all true arithmetical closed (parameter-free) formulas
is arithmetical. Let $\tau(x)$ be the corresponding formula. Also, let us
enumerate all formulas with one parameter x, and let $F_n(x)$ be the
nth formula in this numbering. Consider the formula with a single
parameter x stating that the substitution of the constant x into the
xth formula is false. This formula can be written as follows:

$$\exists z(\neg\tau(z) \wedge \operatorname{Subst}(z, x, x)),$$

where $\operatorname{Subst}(p, q, r)$ is the three-parameter formula that expresses the
property "in a numbering of all parameter-free formulas, p is the
number of the formula obtained by substituting the constant r for
the parameter into the qth one-parameter formula". The property
in quotes describes the graph of a certain computable function (corresponding to simple syntactic operations and the conversion of one
numbering to the other); therefore, it can be expressed by a formula.

Thus we have written a certain formula with a single parameter x.
Suppose that its number is N. We substitute this number N for the
parameter to obtain a parameter-free formula $F_N(N)$. It follows from
the construction that this formula is true if and only if the result of the
substitution of N into the Nth formula (that is, the formula $F_N(N)$
itself!) is false.

The contradiction thus obtained completes the direct proof of
Tarski's Theorem. In this proof, we used the fact that one particular
function (rather than any computable function) was arithmetically
expressible. If you have enough patience, the arithmetical formula
for this function can actually be written. This would make the proof
virtually "tangible".

Now we will present the proof of Gödel's Theorem in the same style. As we have already said, a formal system is a mechanism (algorithm) that can generate certain formulas of the language of arithmetic (for simplicity we assume that only parameter-free formulas are generated). This yields a certain enumerable set, usually specified as the projection of a decidable set. Namely, a notion of *proof* is introduced. Proofs are strings over a certain alphabet. The set of proofs is decidable, that is, we can algorithmically distinguish between proofs and strings that are not proofs. In addition, we have the (decidable) property of a pair of strings x and y that declares x to be a proof of the formula y. Let us enumerate all proofs and formulas and write arithmetical expressions for the decidable properties described above. Thus we arrive at the formula $\text{Proof}(x, y)$ which is true whenever x is a number of the proof of the yth formula.

Now let us write a formula with one parameter x that says that the result of the substitution of x for the parameter into the xth one-parameter formula has no proof:

$$\neg\exists z\exists p[\text{Subst}(z, x, x) \land \text{Proof}(p, z)].$$

Suppose that in the numbering of one-parameter formulas this formula has a number N. Let us substitute N for the parameter. We obtain a formula φ without parameters. By the construction, the formula φ is true whenever the result of the substitution of N into the Nth one-parameter formula is unprovable. But this result is the formula φ itself, so it is true if and only if it is unprovable. It follows that our formal system either allows us to prove the false formula φ (if φ is false; in this case we say that the formal system is not sound) or does not allow us to prove the true formula φ (in this case the formal system is said to be incomplete).

Notice that both proofs we considered resemble the construction of a fixed point as well as the classical Liar's Paradox:

THE STATEMENT IN THE FRAME IS FALSE.

6. Arithmetical hierarchy and the number of quantifier alternations

We have established that the class of arithmetical sets coincides with the union of all the classes Σ_n and Π_n. Let us consider the relationship between the index of a class and the number of quantifier alternations in arithmetical formulas.

To do this, we must explain precisely what formulas are considered as quantifier-free. If addition and multiplication are represented by predicate symbols, then quantifier-free formulas are logical combinations of the formulas $x_i + x_j = x_k$ and $x_i \times x_j = x_k$; here we need a quantifier even to write the formula $a + b + c = d$: it is written as $\exists u((a + b = u) \land (u + c = d))$. However, addition and multiplication are usually thought of as functional symbols, which enables us to use arbitrary polynomials with integer coefficients in quantifier-free formulas. From this point of view, quantifier-free formulas are logical combinations of expressions of the form $P = Q$, where P and Q are arbitrary polynomials with integer coefficients. Then the following theorem holds.

Theorem 73. *A set $A \subset \mathbb{N}$ belongs to the class Σ_n $(n \geq 1)$ if and only if the membership in this set is expressed by a formula in the prenex normal form that starts with the existential quantifier and contains n groups of similar quantifiers.*

(Passing to the complements, we see that a set belongs to the class Π_n if and only if the prenex normal form of the formula for the membership in this set involves n groups of similar quantifiers starting with the universal quantifier.)

Proof. We will only partly prove this theorem. First, if a set is given by a formula in the prenex normal form with n groups of quantifiers, then it belongs to the class Σ_n (or Π_n, depending on which quantifier is the first): recall that the computable bijection between $\mathbb{N}^k$ and $\mathbb{N}$ allows us to replace a group of k consecutive similar quantifiers by one quantifier.

The proof of the converse statement is more difficult. The main problem is the case $n = 1$. In 1970 Yu. V. Matiyasevich, solving the

10th Hilbert Problem, showed that any enumerable set is the set of nonnegative values of a polynomial of several variables (that range over $\mathbb{N}$) with integer coefficients for natural values of the variables. This property can obviously be written in the form of a formula with a quantifier prefix of existential quantifiers. Thus the case of $n = 1$ was explored completely.

Matiyasevich's result makes the proof obvious, since each subsequent class is obtained from the previous one by an additional quantifier.

However, let us try to avoid (rather complicated) Matiyasevich's construction and see what can be extracted from the above proof of the fact that computable functions are arithmetical. Suppose that we are given an arbitrary enumerable set. It can be viewed as the domain of a computable function all of whose values are zero. Then we can apply the procedure described above. It yields a formula that starts with existential quantifiers ("there exist numbers encoding the sequence of values of each variable by means of Gödel's β-function"). Then follows the universal quantifier (over the program step number: "at any step the transition must comply with the program"). Next is a formula that would be quantifier-free if not for the mod-operations it may contain. But the mod-operation can be expressed by universal quantifiers: a statement of the form

$$\forall i \, [\ldots \text{the remainder of } P \text{ when divided by } Q \ldots],$$

where P and Q are some expressions with variables (from the definition of β-encoding), by the definition of the remainder can be rewritten as

$$\forall i \, \forall u \, \forall v \, [(P = uQ + v) \wedge (v < Q)] \Rightarrow [\ldots v \ldots].$$

Therefore, we can represent any enumerable set in the form of a $\exists \ldots \exists \forall \ldots \forall$-formula. This yields a representation of any Σ_n-set in the prenex normal form with $n+1$ groups of quantifiers starting with existential quantifiers, and of any Π_n-set by a formula with $n+1$ groups of quantifiers starting with a universal quantifier. $\qquad \square$

So we get a weaker result (with $n + 1$ instead of n) without referring to the solution of the 10th Hilbert Problem.

Chapter 11

Recursive Functions

1. Primitive recursive functions

Programs with finitely many variables resembled the assembler language; recursive functions, considered in this chapter, are more like functional programming, where some functions are defined in terms of others. We will consider functions with natural arguments and values. In general, these functions are partial, so by an n-ary function we mean a function defined on a subset of $\mathbb{N}^n$ with values in $\mathbb{N}$.

Suppose that we have a k-ary function f and a k-tuple of n-ary functions $g_1, \ldots, g_k$. Then we can form one n-ary function

$$\langle x_1, \ldots, x_n \rangle \mapsto f(g_1(x_1, \ldots, x_n), \ldots, g_k(x_1, \ldots, x_n)).$$

We say that this function is obtained from the functions f and $g_1, \ldots, g_k$ by *substitution*.

Another operation, called *recursion* or *primitive recursion*, is applied to a k-ary function f and $(k+2)$-ary function g. It yields the $(k+1)$-ary function h defined as follows:

$$h(x_1, \ldots, x_k, 0) = f(x_1, \ldots, x_k);$$
$$h(x_1, \ldots, x_k, y+1) = g(x_1, \ldots, x_k, y, h(x_1, \ldots, x_k, y)).$$

Each value in the sequence $h(x_1, \ldots, x_n, 0), h(x_1, \ldots, x_n, 1), \ldots$ is defined in terms of the previous one; therefore, if one of these values is undefined, then all subsequent values are also undefined.

For uniformity, constants are considered as 0-ary functions (functions without arguments); this will allow us to define unary functions recursively.

We say that a function is *primitive recursive* if it can be obtained using substitution and recursion from the following *base* functions: the constant 0, the "successor" function $s\colon x \mapsto x + 1$, and the family of projection functions: for each k, this family consists of k k-ary functions $\pi_k^i(x_1, \ldots, x_k) = x_i$.

Projection functions enable us to perform "nonhomogeneous" substitution. For instance, we can construct the function $\langle x, y \rangle \mapsto f(g(x), h(y, x, y), x)$ from functions f and h by combining them with projections: first we obtain the function $\langle x, y \rangle \mapsto g(x)$ (by substituting π_2^1 into g), then $\langle x, y \rangle \mapsto h(y, x, y)$ (by substituting $\pi_2^2, \pi_2^1, \pi_2^2$ into h), and then the two functions thus obtained together with the function π_2^1 are substituted into f.

Substituting the constant 0 into the add-one function yields the constant (0-ary) function 1. After that we can obtain the constants 2, 3, etc.

2. Examples of primitive recursive functions

Again, it will be helpful to acquire some programming experience, as we did with other computational models.

Addition. The function $\langle x, y \rangle \mapsto \mathrm{sum}(x, y) = x + y$ is obtained by recursion:

$$\mathrm{sum}(x, 0) = x;$$
$$\mathrm{sum}(x, y + 1) = \mathrm{sum}(x, y) + 1.$$

Of course, the right-hand side of the second equation must be represented in terms of substitution. Formally, here the function $h(x, y, z)$ from the definition of recursion must be set equal to $s(z)$, where s is the successor function.

Multiplication. The function $\langle x, y \rangle \mapsto \mathrm{prod}(x, y) = xy$ is obtained by recursion (that involves addition):

$$\mathrm{prod}(x, 0) = 0;$$
$$\mathrm{prod}(x, y + 1) = \mathrm{prod}(x, y) + x.$$

Likewise, we can pass from multiplication to exponentiation.

Truncated subtraction. We consider "truncated subtraction" $x \mathbin{\dot-} y = x - y$ for $x \geq y$ and $x \mathbin{\dot-} y = 0$ for $x < y$, because we deal only with natural (nonnegative integer) numbers. The unary function of truncated subtraction of 1 is defined recursively:

$$0 \mathbin{\dot-} 1 = 0;$$
$$(y + 1) \mathbin{\dot-} 1 = y.$$

(This is only a formal recursion, since the previous value is not used.) Then the truncated subtraction for arbitrary arguments can be defined as follows:

$$x \mathbin{\dot-} 0 = x;$$
$$x \mathbin{\dot-} (y + 1) = (x \mathbin{\dot-} y) \mathbin{\dot-} 1.$$

3. Primitive recursive sets

A set is called *primitive recursive* if its characteristic function is primitive recursive. (Or, equivalently, if it is the set of zeros of a primitive recursive function. The equivalence of the two versions of the definition follows by substitution into the function $x \mapsto 1 \mathbin{\dot-} x$.)

The intersection and union of primitive recursive sets are primitive recursive (we can add or multiply the two functions whose sets of zeros are the sets in question). The complement of a primitive recursive set is primitive recursive. Identifying sets and properties, we can say that conjunctions, disjunctions, and negations of primitive recursive properties are primitive recursive.

The properties $x = y$ and $x \neq y$ are primitive recursive ($x = y$ if and only if $(x \mathbin{\dot-} y) + (y \mathbin{\dot-} x) = 0$).

The function $f(x)$ defined by the relation

$$f(x) = [\ \textbf{if } R(x) \textbf{ then } g(x) \textbf{ else } h(x) \textbf{ fi }]$$

is primitive recursive if the functions g and h and the property R are also. Indeed, $f(x)$ can be written as $r(x)g(x) + (1 \mathbin{\dot-} r(x))h(x)$, where r is the characteristic function of the property R.

Now we can write a formula that adds one modulo n to a number less than n:

$$x + 1 \bmod n = [\ \mathbf{if}\ x + 1 = n\ \mathbf{then}\ 0\ \mathbf{else}\ x + 1\ \mathbf{fi}\ ].$$

After that the function $x \bmod n$ (the remainder upon division by n) can be defined recursively:

$$0 \bmod n = 0;$$

$$(x + 1) \bmod n = (x \bmod n) + 1 \bmod n.$$

Let us show that bounded quantifiers applied to primitively recursive properties (sets) yield primitive recursive properties again. This means, for instance, that if a property $R(x, y)$ is primitive recursive, then the properties

$$S(x, z) = (\exists y \leq z)\, R(x, y)$$

and

$$T(y, z) = (\forall y \leq z)\, R(x, y)$$

are primitive recursive as well. To prove this, we notice that in terms of functions, a bounded quantifier can be replaced by multiplication or summation: if a property $R(x, y)$ is equivalent to $r(x, y) = 0$, then

$$S(x, z) \Leftrightarrow \left[\prod_{y=0}^{z} r(x, y) = 0\right].$$

The product here can be defined recursively:

$$\prod_{y=0}^{0} r(x, y) = r(x, 0);$$

$$\prod_{y=0}^{t+1} r(x, y) = \left[\prod_{y=0}^{t} r(x, y)\right] \cdot r(x, t + 1);$$

summation is handled similarly.

Now it is readily seen that the property "to be a prime number" is primitive recursive (any smaller number is either zero or 1 or is not a divisor).

We proceed to show that if the graph of a function f is primitive recursive and the values of f are bounded from above by a primitive recursive function g, then the function f itself is primitive recursive. Indeed, if r is the characteristic function of the graph, i.e., $r(x, y) = 1$ for $y = f(x)$ and $r(x, y) = 0$ for $y \neq f(x)$ (for simplicity we consider only the case of unary functions), then

$$f(x) = \sum_{i=0}^{\infty} y \cdot r(x, y);$$

in fact, the summation is bounded from above by the value $g(x)$. It remains to use the primitive recursiveness of a bounded sum.

Hence it is not difficult to derive that if a function g and a property $R(x, y)$ are primitive recursive, then the function f defined by the following formula is also primitive recursive:

$$x \mapsto f(x) = \text{the smallest } y \leq g(x) \text{ such that } R(x, y)$$

(if there is no such y for a given x, we set the value of f equal to, say, $g(x) + 1$). Indeed, the graph of the function f can be specified using bounded quantifiers.

This function-building operation is called *bounded minimization* as distinct from unbounded minimization, in which no specific boundary $g(x)$ is known in advance. As we will see, the function created in the second case need not be primitive recursive.

Bounded minimization can be used to show that the function $x \mapsto$ (the smallest prime number greater than x) is primitive recursive (Euclid's proof of the infinity of the set of prime numbers yields the boundary function $g(x) = x! + 1$, and the factorial is primitive recursive). Now it is easy to recursively define the function $n \mapsto$ (the nth prime).

4. Other forms of recursion

Thus far, by "recursive definition of a function" we meant the definitions that employed primitive recursion. However, this term can be understood in a wider sense, as any definition of a function that relates its value at a given point with its other values. As we will see in the discussion of Ackermann's function below, some of these schemes of recursive definitions lead us beyond the class of primitive recursive functions. On the other hand, some of them can be reduced to the scheme considered above.

We will give two examples of the latter: simultaneous definition of several functions and recursion that involves arbitrary numbers smaller than the value of the variable.

Joint recursion. Let f and g be two unary functions specified by the relations

$$f(0) = a,$$
$$g(0) = b,$$
$$f(n+1) = F(n, f(n), g(n)),$$
$$g(n+1) = G(n, f(n), g(n)),$$

where a and b are arbitrary numbers and F and G are primitive recursive ternary functions. Then the functions f and g are primitive recursive.

To prove this we will need a primitive recursive numbering of pairs, i.e., a function $\langle x, y \rangle \rightarrow [x, y]$ (where the brackets denote the number of a pair) which is primitive recursive along with its two inverse functions p_1 and p_2 (that return the first and second components of a pair given its number). Using this numbering, we can recursively define the function $h(n) = [f(n), g(n)]$:

$$h(0) = [a, b],$$
$$h(n+1) = [F(n, p_1(h(n)), p_2(h(n))), G(n, p_1(h(n)), p_2(h(n)))].$$

If h is a primitive recursive function, then the functions f and g (obtained by substituting the function h into p_1 and p_2) are primitive recursive as well.

It remains to construct a primitive recursive numbering of pairs. The desired bijection $\mathbb{N} \times \mathbb{N} \rightarrow \mathbb{N}$ is illustrated by the following table:

```
6
3  7
1  4  8
0  2  5  9
```

Problem 86. Show that this bijection can be specified by a quadratic polynomial of two variables. (*Hint*: this polynomial is uniquely determined by the values shown in the table.)

The primitive recursiveness of the inverse mappings p_1 and p_2 is established by bounded minimization, since $p_1(n)$ is the minimal $x \leq n$ such that there exists a $y \leq n$ with $[x, y] = n$.

A less symmetric numbering of pairs is given by the formula $[a, b] = (2a + 1)2^b$. We may also notice that it is not necessary for each nonnegative integer to be the number of some pair; so another suitable numbering is $[a, b] = 2^a 3^b$.

A similar construction is applicable in the case of more than two simultaneously defined functions and for functions of many variables.

Using previous values. The next statement shows that a recursive definition can involve not only the value at the immediately preceding point, but any previous values as well.

Theorem 74. *Suppose that g is a primitive recursive unary function such that $g(x) < x$ for $x > 0$, F is a primitive recursive binary function, and c is an arbitrary constant. Then the function h defined by the relations*

$$h(0) = c,$$
$$h(x) = F(x, h(g(x))) \text{ for } x > 0$$

is primitive recursive.

Proof. We will prove this theorem using the following numbering of finite sequences of natural numbers: the empty sequence is given the number 1; the one element sequence $\langle a \rangle$ is given the number 2^{a+1}; the sequence $\langle a, b \rangle$ is given the number $2^{a+1}3^{b+1}$; the sequence $\langle a, b, c \rangle$ is given the number $2^{a+1}3^{b+1}5^{c+1}$, and so on (the bases of the powers are successive primes). Let us denote the number of the sequence $\langle a, b, \ldots, z \rangle$ by $[a, b, \ldots, z]$. In a certain sense, this numbering is primitive recursive. Of course, this statement should not be understood literally, because this numbering is a "function with a variable number of arguments". But many related functions are primitive recursive. In particular, such are the functions

- Length(x) = the length of the sequence number x;

- Select(i, x) = the ith term of the sequence number x;
- Append(x, y) = the number of the sequence obtained by appending y to the sequence number x.

All these functions (as well as many similar ones) can be constructed using various operations with prime numbers and factors, and these operations have already been considered.

Now it will suffice to show that the function

$$x \mapsto H(x) = [h(0), h(1), \ldots, h(x)]$$

is primitive recursive. We have $H(0) = [c]$ and

$$H(k + 1) = \text{Append}(H(k), F(k + 1, \text{Select}(g(k + 1), H(k)))).$$

$\square$

A similar approach works for the recursive definitions that use several previous values.

5. Turing machines and primitive recursive functions

We have considered various techniques for building primitive recursive functions. However, it is not quite clear yet how wide this class is. Now we will show that it includes all functions that are computable in a reasonable (not extremely long) time.

Theorem 75. *Any function computable by a Turing machine whose computation time is bounded by a primitive recursive function of the input length is primitive recursive.*

Proof. By our definition of a Turing machine, its input and output are strings of zeros and ones. Since the arguments and values of primitive recursive functions are numbers, the theorem will make sense only if we agree to identify numbers and strings. As we have already said, a number n is identified with the string obtained by erasing the leading bit 1 from the binary notation of the number $n + 1$.

To simulate a Turing machine by a suitable program, we encoded the machine configurations by four numbers corresponding to the codes of the left-hand and right-hand parts of the tape, the current state and the character under the head. In our encoding system the left-hand part of the tape was interpreted as the notation of some number to the base equal to the tape alphabet size (the blank symbol being read as zero) and the right-hand part of the tape was treated likewise, but in reverse order (low bits starting from the head). Under this agreement, inserting or deleting the character amounts to simple arithmetical operations (deleting is division and inserting is multiplication by the base of the number system followed by addition). In this encoding, the transition functions (four functions of four variables that yield the next configuration of the machine as a function of the preceding one) are expressed by simple formulas and are primitive recursive.

Now let us consider the iterated transition function, which specifies the Turing machine configuration after t steps. More exactly, we have four functions of five variables (the first four variables encode the state and the fifth is the number of steps). They are defined by joint recursion, considered in the previous section. Therefore, these functions are primitive recursive. We will assume that once the machine terminates, its configuration never changes again. Since the number of steps of the Turing machine is bounded by a primitive recursive function, it will suffice to substitute this bound for the fifth argument (the number of steps) to ascertain that the terminal configuration of the machine is a primitive recursive function of its initial configuration. Hence the output of the machine is a primitive recursive function of the input.

This reasoning implicitly involves primitive recursiveness of various functions used in the conversions from one data representation to another. For instance, the input of the Turing machine is a binary string, which we have agreed to identify with a certain number x. This input is converted into the initial configuration of the Turing machine, which is encoded by four numbers. It is important that these four numbers primitive-recursively depend on x. This becomes clear after we notice that the conversion is connected with transition from one number notation to another (the same string encodes different numbers in different notations), and the primitive recursiveness of this sort of functions is readily established using our technique. In addition, we must compute the input length (in order to substitute it into the primitive recursive function that bounds the number of steps), and in the end the result must be primitive-recursively extracted from the output configuration and recoded. But all these operations remain within the scope of the above-considered methods, and we will not dwell on them.   □

This theorem convinces us of the primitive recursiveness of many pretty sophisticated functions. For instance, consider the function $n \mapsto$ (the nth decimal digit of the number π). It is well known that millions of these digits have been calculated. So there is good reason to believe that there exist rather efficient algorithms to compute π: even taking into account all the clumsiness of Turing machines from the programmers' viewpoint, it would be very strange if the computation of the nth digit of π needed a time greater than, say, $c \times 2^n$ for a large enough c. But this estimate is primitive recursive, which allows us to refer to our theorem. (In fact, we have a wide safety margin here: there exist primitive recursive functions that grow much faster than 2^n.)

6. Partial recursive functions

Operations of primitive recursion and substitution do not take us beyond the class of total functions. The situation is different with the *minimization*, which has already been mentioned. It is applied to a $(k+1)$-ary function f and yields a k-ary function g defined as follows: $g(x_1, \ldots, x_k)$ is *the smallest y for which $f(x_1, \ldots, x_k, y) = 0$.*

The meaning of the italicized words is clear if the function f is total. Otherwise, they must be understood as follows: the value $g(x_1, \ldots, x_k)$ is equal to y if $f(x_1, \ldots, x_k, y)$ is defined and equal to zero, and all the values $f(x_1, \ldots, y_k, y')$ for $y' < y$ are defined and not equal to zero.

Minimization is often denoted by

$$g(x_1, \ldots, x_k) = \mu y \, (f(x_1, \ldots, x_k, y) = 0),$$

and for this reason it is also called the μ-*operator*.

Clearly, this definition ensures the computability of g whenever f is computable (we search through all y in ascending order waiting for the zero value to appear).

Problem 87. Show that if the definition is modified so as to permit $f(x_1, \ldots, x_k, y')$ to be undefined for $y' < y$, then the function g can be noncomputable even if f is computable.

The functions that can be built from the base functions (zero, projection, and the add-one function) by means of substitution, primitive recursion, and minimization are called *partial recursive*. If such a function turns out to be total, then it is called a *total* (or *general*) *recursive* function.

Theorem 76. *Any Turing machine computable function is partial recursive.*

Proof. Let f be a unary function computable by a Turing machine M. Denote by $T(x, y, t)$ the property of the machine M to return y on the input x in a time not exceeding t. As we have seen above, from the input of the Turing machine and the time t, we can primitive-recursively compute its state at the moment t. It is also clear that we can find out whether the machine has completed the job and, if it has, whether the answer was equal to y. Thus the property T is primitive recursive.

Now let us combine the arguments y and t into a pair by a primitive recursive numbering; we obtain a primitive recursive function T' such that $T'(x, [y, t]) = T(x, y, t)$. Now we can write $f(x) = p_1(\mu z T'(x, z))$, where p_1 assigns the first component of a pair to the pair's number, and μz stands for "the smallest z such that $\ldots$". Thus the function f is partial recursive. $\qquad\square$

The converse statement is also true:

Theorem 77. *Any partial recursive function is Turing machine computable.*

Proof. It is not hard to write a program with finitely many variables computing any partial recursive function (substitution is the consecutive execution of programs, recursion is a `for`-loop, minimization is

a while-loop; both types of loops are easily implemented by means
of conditional goto-statements).

After that it remains to refer to the fact that any function computable by a program with finitely many registers is computable by a
Turing machine (as we have shown in Section 10.2, Theorem 66).   □

Therefore, if we believe in the "Turing Thesis", which claims that
any computable function is Turing machine computable, then we must
just as well believe in the "Church Thesis" (any computable function
is partial recursive), so these theses are equivalent.

The real history of the subject is more complicated and can roughly be
described as follows. The definition of a primitive recursive function was
given by the great logician Kurt Gödel and was used as a technical means
in the proof of Gödel's Incompleteness Theorem in the early 1930s. The
definition of a total recursive function was also given by Gödel (it differs
from, but is equivalent to, the one we gave above). The American logician
Alonzo Church formulated his thesis for total functions, having conjectured
that any total computable function is total recursive. Then the American
mathematician Kleene suggested that this thesis should be extended to
partial functions.

At the same time, the English mathematician Turing and the American
mathematician Post proposed their models of abstract computing machines
(Turing and Post machines) that differed only in some details, and conjectured that these machines cover the entire class of algorithmic processes.
Soon it became clear that the computability of functions by these machines
is equivalent to partial recursiveness. (More historical details can be found
in Kleene's book [5].)

Now the expressions "Turing Thesis", "Church Thesis", "Post Thesis",
and so on are usually used synonymously: these theses claim that the
class of computable (in the intuitive sense) partial functions coincides with
the class of partial recursive functions (the Church Thesis), or Turing-
computable functions (Turing Thesis), and so on. All these versions are
equivalent, because all usual formal definitions of computability (partial
recursiveness, Turing machines, etc.) lead to the same class of functions.

(Let us notice parenthetically that another computational model, normal algorithms or Markov algorithms, were proposed by Andrei Andreevich
Markov Jr. (Markov chains and Markov processes are named after his fa-
ther, A. A. Markov Sr.). But this happened later, in the 1950s. The
corresponding principle (any algorithm is equivalent to a normal one) was
introduced by Markov; he called it the normalization principle. In Markov's

papers, normal algorithms were used for the construction of an undecidable string-rewriting system (see Section 9.4). It is worth mentioning that Markov has explicitly written in every detail the construction of the universal algorithm and gave an exact proof of its correctness; it seems that this achievement has never been repeated since then: no one was persistent enough to write a compiler of some programming language in this language and formally prove its correctness.)

Our proofs of Theorems 76 and 77 yield one more corollary, sometimes called *Kleene's Normal Form Theorem*:

Theorem 78. *Any partial recursive function f is representable in the form*

$$f(x) = a(\mu z(b(x, z) = 0)),$$

where a and b are primitive recursive functions.

Proof. Indeed, any partial recursive function is Turing machine computable, and hence, as is seen from the proof of Theorem 76, can be represented in the desired form (where a is the function that assigns the first component of a pair to the pair's number). $\square$

This theorem can be extended from the unary functions f to a similar statement for functions of several variables (the proof remains almost the same).

Problem 88. Show that one μ-operator applied last will not suffice: a partial recursive function may not be representable in the form

$$f(x) = \mu z(b(x, z) = 0)$$

where b is a primitive recursive function.

Kleene's Normal Form Theorem implies the following statement:

Theorem 79. *Any enumerable set is the projection of a primitive recursive set.*

Proof. Any enumerable set is the domain of a recursive function. Having represented the function in normal form, we see that its domain is the projection of the set $\{\langle x, z\rangle \mid b(x, z) = 0\}$. $\square$

7. Oracle computability

The definition of the class of partial recursive functions is readily modified for the case of oracle computability. Suppose that α is a total function. Consider the class $\mathcal{F}[\alpha]$ consisting of the base functions, the function α, and all the functions that can be obtained from them by substitution, primitive recursion, and minimization.

(Formally, $\mathcal{F}[\alpha]$ is the minimal class containing the base functions and the function α which is closed under substitution, primitive recursion, and minimization. This minimal class exists: it suffices to take the intersection of all the classes with these properties.)

Theorem 80. *The class $\mathcal{F}[\alpha]$ consists of all α-computable functions (that is, functions computable by programs that call α as an oracle).*

Proof. First of all, let us notice that all functions of the class $\mathcal{F}[\alpha]$ are computable by the programs in question. This can be explained, for example, as follows. Programs with finitely many variables compute all partial recursive functions. If we are allowed to use the statement $\mathbf{a} := \alpha(\mathbf{b})$, then they will just as well compute all functions of the class $\mathcal{F}[\alpha]$.

The converse statement is more interesting: we want to prove that if a certain (in general, partial) function is α-computable, then it can be constructed from the base functions and the function α by substitution, recursion, and minimization.

To this end, let us recall the criterion of relative computability from Chapter 7.2 (Theorem 45). Let a function f be computable relative to a total function α. Then, as we know, there exists an enumerable set W of triples of the form $\langle x, y, t \rangle$, where x and y are natural numbers, and t is a pattern (a function with finite domain), which is consistent (the patterns of any two triples with the same x, but different y, are not coherent) and satisfies

$$f(x) = y \Leftrightarrow \exists t \, (\langle x, y, t \rangle \in W \text{ and } t \text{ is a part of } \alpha).$$

We will show that the property "t is a part of α" is primitive recursive relative to α, i.e., its characteristic function is obtained from the base functions and from α by means of substitution and recursion operations. (Recall that we identify patterns with their numbers in a certain numbering; the choice of the numbering is discussed later.)

Then it will suffice to write W as the projection of a primitive recursive set $(\langle x, y, t \rangle \in W \Leftrightarrow \exists u \, (v(x, y, t, u) = 0)$, where v is a primitive recursive function), and use the equation

$$f(x) = p_1(\mu z \; v'(x, z) = 0),$$

where v' is an α-primitive recursive function such that $v'(x, [y, t, u]) = 0$ if and only if $[v(x, y, t, u) = 0$ and t is a part of the function $\alpha]$, and the function p_1 computes the first component y of the triple $\langle y, t, u \rangle$ from the triple's number $[y, t, u]$.

Finally, it remains to show that the set $\{t \mid$ the pattern with number t is a part of $\alpha\}$ is α-primitive recursive. In the proof we will assume that the numbering of patterns is chosen so that the following functions are primitive recursive (they are well defined for patterns with nonempty domain; for the empty pattern we define them at will):

- last-x(t), the largest of the numbers on which the pattern with number t is defined;

- last-y(t), the value of the pattern with number t at the maximal point in its domain (recall that patterns are functions);

- all-but-last(t), the number of the pattern obtained from the pattern with number t by removing the maximal point of the domain.

Then we can write the following recursive definition: the pattern with number t is a part of the function α if either this pattern is empty or $\alpha(\text{last-x}(t)) = \text{last-y}(t)$ and the pattern with number all-but-last(t) is a part of the function α.

This definition uses the type of recursive definition considered in Theorem 74 (p. 145); the value of the function is defined recursively in terms of its values at smaller points. We only have to choose the numbering of patterns so as to ensure that all-but-last(t) is smaller than t for all t. But this is easy: for instance, we can use primes, assigning to the pattern $\{\langle a, b \rangle, \ldots, \langle e, f \rangle\}$ the number $p_a^{(b+1)} \cdots p_e^{(f+1)}$, where p_i is the ith prime number (so that $p_0 = 2$, $p_1 = 3$, $p_2 = 5, \ldots$). $\quad\square$

Notice that the proof could be slightly simplified by using only patterns whose domain is an initial segment of the set of natural numbers. In that case we could start with establishing the primitive recursiveness (relative to α) of the function $n \mapsto [\alpha(0), \alpha(1), \ldots, \alpha(n)]$. Also, it is easily seen that in the definition of relative computability, we can confine ourselves only to patterns of this type.

8. Estimates of growth rate. Ackermann's function

Now we turn to the question that could have been asked long ago: do their exist total recursive, but not primitive recursive, functions? We will give two proofs of their existence. The first of them is based on general considerations:

Theorem 81. *There exists a binary total computable function universal for the class of all unary primitive recursive functions.*

Obviously, if U is such a function, then the function d defined by $d(n) = U(n, n) + 1$ will be total, computable, and distinct from any primitive recursive function (it differs from the nth function at the point n).

Proof. Any primitive recursive function is built from the base functions by a sequence of substitutions and recursions. This sequence can be represented as a string over a finite alphabet, which can be viewed as a sort of a program (that successively defines various primitive recursive functions by specifying for each of them other functions they are built from and the operations applied). From all these programs, we select programs of unary functions (of course, intermediate functions in the programs can have any number of variables). The set of all such programs is decidable, and they can be computably numbered. Then the function $\langle n, x \rangle \mapsto$ (the result of application of the function specified by the nth program to the number x) is computable and, by construction, is universal for the class of primitive recursive functions. $\square$

However, it is interesting to point out a more specific reason that prevents some computable functions from being primitive recursive.

One such reason is that primitive recursive functions cannot grow too fast. This idea dates back to Ackermann who constructed a function growing faster than all primitive recursive functions called *Ackermann's function*. We will give a slightly different construction based on the same idea.

First, we define a sequence $\alpha_0, \alpha_1, \ldots$ of unary total functions. Denote by $f^{[n]}(x)$ the iteration $f(f(\ldots f(x)\ldots))$, where the function f is repeated n times. Set $\alpha_0(x) = x + 1$ and

$$\alpha_i(x) = \alpha_{i-1}^{[x+2]}(x)$$

(we will explain later why it is convenient to apply the function α_{i-1} exactly $x + 2$ times). For example, $\alpha_1(x) = \alpha_0^{[x+2]}(x) = 2x + 2$.

The following properties are obvious (a formal proof is by induction):

- $\alpha_i(x) > x$ for all i and x;
- $\alpha_i(x)$ is a strictly increasing function of x;
- $\alpha_i(x)$ is a strictly increasing function of i (for any fixed x);
- $\alpha_i(x) \geq \alpha_{i-1}(\alpha_{i-1}(x))$.

Now we can estimate the growth rate of any primitive recursive function.

Theorem 82. *Let f be a primitive recursive function of n variables. Then for a certain number k,*

$$f(x_1, \ldots, x_n) \leq \alpha_k(\max(x_1, \ldots, x_n))$$

for all $x_1, \ldots, x_n$.

Proof. The idea is simple: we will bound the growth rate of the composition of functions if bounds for all involved functions are given; the same will be done for recursion. Formally, the proof follows the inductive definition of primitive recursive functions.

For the base functions, the bound is obvious. Consider a function obtained by substitution. Suppose that

$$f(x) = g(h_1(x), \ldots, h_k(x))$$

(here the letter x stands for the tuple of variables). If all the functions $h_1, \ldots, h_k$ and the function g are bounded by α_N, i.e., $h_i(x) \leq \alpha_N(\max(x))$ for all i and x and $g(y) \leq \alpha_N(\max(y))$ (here $\max(u)$ means the greatest element in the tuple u), then $f(x)$ does not exceed

$$\alpha_N(\max(h_1(x), \ldots, h_k(x))) \leq \alpha_N(\alpha_N(x)) \leq \alpha_{N+1}(x)$$

(we use the above-mentioned properties of the functions α_i).

Recursion is treated in a similar (but somewhat more difficult) way. Suppose that a function f is defined recursively:

$$f(x, 0) = g(x);$$
$$f(x, n+1) = h(x, n, f(x, n)),$$

where x denotes the tuple of variables. If the functions g and h are bounded by the function α_N, then

$$f(x, 1) = h(x, 0, f(x, 0)) \leq \alpha_N(\max(x, 0, f(x, 0)))$$
$$\leq \alpha_N(\max(x, 0, \alpha_N(\max(x)))) \leq \alpha_N(\alpha_N(\max(x)))$$

(the last inequality follows from the estimate $\alpha_N(t) > t$). Similarly, $f(x, 2) \leq \alpha_N(\alpha_N(\alpha_N(\max(x))))$ and, in general,

$$f(x, i) \leq \alpha_N^{[i+1]}(\max(x)) \leq \alpha_{N+1}(\max(i, \max(x))),$$

which completes the proof. $\square$

Notice that each application of substitution or recursion increases the index i in the upper bound α_i by 1, so the function specified by at most 100 operators is bounded by α_{101}.

The following statement is an obvious corollary of the above estimate:

Theorem 83. *The function $A(n) = \alpha_n(n)$ grows faster than any primitive recursive function.*

It should be mentioned that the definition of Ackermann's function (more exactly, of the function $\langle n, x \rangle \mapsto \alpha_n(x)$) can be considered a recursive definition: each value of this function is defined in terms of others values of the same function, with a smaller first argument.

This is an example of a recursive definition that cannot be reduced to primitive recursion.

Problem 89. Show that the direct enumeration (in ascending order) of an infinite primitive recursive set need not be primitive recursive.

Problem 90. Show that the function inverse to a primitive recursive bijection $i\colon \mathbb{N} \to \mathbb{N}$ need not be primitive recursive.

Problem 91. Prove that if g is a ternary primitive recursive function and h is unary primitive recursive function, then the binary function f defined by the equations

$$f(x,0) = h(x),$$
$$f(x,i+1) = g(x,i,f(2x,i))$$

is primitive recursive.

Bibliography

[1] J. Barwise (Ed.), *The handbook of mathematical logic*. North-Holland, Amsterdam, 1977.

[2] George S. Boolos, John P. Burgess, and Richard C. Jeffrey, *Computability and logic*. Fourth edition, Cambridge Univ. Press, Cambridge, 2002.

[3] Nigel Cutland, *Computability: An introduction to recursive function theory*. Cambridge Univ. Press, Cambridge, 1980.

[4] Yu. L. Ershov, *Theory of numberings*. Nauka, Moscow, 1977. (Russian)

[5] S. K. Kleene, *Introduction to metamathematics*. (11th Printing) North-Holland, Amsterdam, 1996.

[6] A. I. Mal'tsev, *Algorithms and recursive functions*. Nauka, Moscow, 1965. (Russian)

[7] Yu. I. Manin, *Computable and incomputable*. Sov. Radio, Moscow, 1980. (Russian)

[8] ———, *A course in mathematical logic*. Springer-Verlag, New York, 1977.

[9] M. Minsky, *Computation: Finite and infinite machines*. Prentice-Hall, Englewood Cliffs, NJ, 1967.

[10] Piergiorgio Odifreddi, *Classical resursion theory*. Vols. I, II, North-Holland, Englewood Cliffs, NJ, 1989, 1999.

[11] Hartley Rogers, Jr., *Theory of recursive functions and effective computability*. MIT Press, Cambridge, MA, 1987.

[12] A. Shen and N. K. Vereshchagin, *Basic set theory*, Amer. Math. Soc., Providence, RI, 2002.

[13] J. Shoenfield, *Degrees of undecidability*. North-Holland, Englewood Cliffs, NJ, 1971.

[14] Michael Sipser, *Introduction to the theory of computation*, PWS, Boston, 1997.

[15] Robert I. Soare. *Recursively enumerable sets and degrees.* Springer-Verlag, Berlin, 1987.

[16] V. A. Uspensky, *Lectures on computable functions.* Fizmatgiz, Moscow, 1960. (Russian)

[17] V. A. Uspensky and A. L. Semenov, *Algorithms: Main ideas and applications.* Kluwer, Dordrecht, 1993.

Glossary

Wilhelm ACKERMANN, Mar. 29, 1896, Schoenebeck [Kr. Altena] (Germany) – Dec. 24, 1962, Luedenscheid (Germany)   144, 155

Alonzo CHURCH, June 14, 1903, Washington, D.C. (USA) – Aug. 11, 1995, Hudson, Ohio (USA)   150

EUCLID of Alexandria, about 325 (?) BC – about 265 (?) BC, Alexandria (now Egypt)   112, 143

Pierre de FERMAT Aug. 17, 1601, Beaumont-de-Lomagne (France) – Jan. 12, 1665, Castres (France)   9

Richard FRIEDBERG   85

Kurt GÖDEL, Apr. 28, 1906, Brünn, Austria-Hungary (now Brno, Czech Republic) – Jan. 14, 1978, Princeton, New Jersey (USA)   130, 137, 150

David HILBERT, Jan. 23, 1862, Königsberg, Prussia (now Kaliningrad, Russia) – Feb. 14, 1943, Göttingen (Germany)   136

Stephen Cole KLEENE, Jan. 5, 1909, Hartford, Connecticut (USA) – Jan. 25, 1994, Madison, Wisconsin (USA)   150, 151

Donald KNUTH, born Jan. 10, 1938, Milwaukee, Wisconsin (USA)   112

Andrei Andreevich MARKOV, Jr., Sep. 9, 1903 – Oct. 11, 1979, Moscow (USSR)   122, 150

Andrei Andreevich MARKOV, Sr., June 14, 1856, Ryazan (Russia) –
July 20, 1922, Petrograd (now St. Petersburg, Russia)   150

Yuri Vladimirovich MATIYASEVICH, born Mar. 2, 1947, Leningrad
(USSR)   9, 136

Albert Abramovich MUCHNIK, born Jan. 2, 1934, Moscow (USSR)
85

Petr Sergeevich NOVIKOV, Aug. 28, 1901, Moscow (Russia) – Jan.
9, 1975, Moscow (USSR)   122

Emil Leon POST, Feb. 11, 1897, Augustów (Poland) – Apr. 21, 1954,
New York (USA)   ix, 7, 17, 85, 122, 150

Henry Gordon RICE   103

Hartley ROGERS, Jr.   34, 53

Michael SIPSER   56

Alfred TARSKI, Jan. 14, 1902, Warsaw (Russia, now Poland) – Oct.
26, 1983, Berkeley, California (USA)   133, 134

Axel THUE, Feb. 19, 1863, Tönsberg (Norway) – Mar. 7, 1922, Oslo
(Norway)   113

Alan Mathison TURING, June 23, 1912, London (England) – June 7,
1954, Wilmslow, Cheshire (England)   ix, 72, 107, 146, 149, 150

Vladimir Andreevich USPENSKII, born Nov. 27, 1930   103

Index

Titles in This Series